ASSESSMENT AND PROFILING IN SCIENCE

For Fanou

Assessment and Profiling in Science:

A Practical Guide

Chris Christofi

CASSELL

Cassell Educational Limited
Artillery House
Artillery Row
London SW1P 1RT

British Library Cataloguing in Publication Data

Christofi, Chris
  Assessment and profiling in science.
  1. Great Britain. Secondary schools.
  Curriculum subjects : Science. Academic
  achievement of students. Assessment
  I. Title
  507′.6

ISBN 0 304 31554 0

Phototypesetting by Inforum Ltd, Portsmouth
Printed and bound in Great Britain by Biddles Ltd, Guildford

Contents

Acknowledgements

First and foremost I should like to thank Bob Emmerson for his time and valuable comments on the rough draft; also many thanks to Robert Taylor for his guidance over the writing of the section on the Assessment of Performance Unit.

I should also like to thank those who provided me with materials or gave of their time. Thanks to Les Cowling, Diane Fairbairn, David Garforth, Roger Murphy, Harry Torrance and those, often unknown to me, who so willingly sent material to me through the post.

I should also like to thank my wife, Janet, for her patience and support.

In addition to these people I should like to express my gratitude to the following for permission to reproduce material:
Assessment of Performance Unit
Bell and Hyman Publishers
Cambridge Local Examinations Syndicate
Cassell Publishers
City and Guilds of London Institute
Heinemann Educational Books
Inner London Education Authority
Joint Matriculation Board
Midland Examining Group
Oxford Certificate of Educational Achievement
The Associated Examining Board
University of Oxford Delegacy of Local Examinations.

Abbreviations used in this text

AEB   – Associated Examining Board
AMMA   – Assistant Masters and Mistresses Association
APU   – Assessment of Performance Unit
BTEC   – Business and Technician Education Council Courses
CGLI   – City and Guilds of London Institute
CPVE   – Certificate of Pre-Vocational Education
CSE   – Certificate of Secondary Education
DES   – Department of Education and Science
FEU   – Further Education Unit (DES)
FHE   – Further and Higher Education
GCE   – General Certificate of Education
GCSE   – General Certificate of Secondary Education
GRC   – Grade Related Criteria
HMI   – Her Majesty's Inspectors
ILEA   – Inner London Education Authority
INSET   – In-Service Training
IPM   – Institute of Personnel Management
JMB   – Joint Matriculation Board
LEA   – Local Education Authority
MEG   – Midland Examining Group
MSC   – Manpower Services Commission
NAS/UWT – National Association of Schoolmasters/Union of Women Teachers
NEA   – Northern Examining Association
NUT   – National Union of Teachers
OCEA   – Oxford Certificate of Educational Achievement

PAR	– Personal Achievement Record
PPR	– Pupil Personal Record
RPA	– Record of Personal Achievement
RPE	– Record of Personal Experience
RSA	– Royal Society of Arts
SCRE	– Scottish Council for Research in Education
SED	– Scottish Education Department
SEG	– Southern Examining Group
SREB	– Southern Regional Examinations Board
TAPS	– Techniques for the Assessment of Practical Skills in Foundation Science
TIP	– Teachers Item Planner
TVEI	– Technical and Vocational Education Initiative

Preface

In recent years, there has been much progress in methods of recording and also a move towards recording a much wider variety of abilities. In science, especially with the advent of the General Certificate of Secondary Education (GCSE), this has meant greater internal practical assessment.

This book has been written with the ethos of the present assessment movement in mind, that is that assessment is for differentiation (what the students know, understand and can do) rather than discrimination.

It is my intention to discuss a variety of assessments that may be used to gather information for profiling and to further discuss the types of profiles that can be produced. Although the main aim is to put emphasis, particularly in Chapters 2 and 3, on science education, assessing and profiling do cross curricular boundaries. Similar information is sought for in different subjects and is recorded in similar ways; therefore what is said about science assessment and profiling may also apply to other subjects, to vocational training, to technical training, etc.

In Chapter 1, the emphasis is on an in-depth attempt to define profiling and there is little more than an outline of what is meant by assessment. The reason for this is that profiling is a relatively new activity in education and its definition is still evolving, whereas educational assessment is now well established in schools and its meaning causes little, if any, confusion.

Chapters 2 and 3 are intended as an in-depth study of assessment; a general discussion of the topic in Chapter 2 is followed by a consideration of the issues affecting science, with especial reference to the internal assessment of practical skills. The information obtained from assessments needs to be organised and recorded and Chapter 3 looks at schemes that are endeavouring to do this in science.

A discussion of current practices in recording and profiling takes place in Chapter 4, with particular reference to the main schemes being used in England and Wales. Accurate and relevant assessment and well-constructed profiles are interdependent. Poor practice in one area negates the effect of the other.

One cannot discuss such major educational themes without considering their implications for the curriculum. In Chapter 5, some consideration is given to how various areas of the curriculum may be affected, such as subject matter, resources, syllabus content, crosscurricular co-operation, GCSE, Technical and Vocational Educational Initiative (TVEI) and other current educational initiatives.

There is no attempt to provide a balance in the discussion of assessment and profiling, but instead, the most pertinent points from each area are highlighted.

Chapter 1

What Is Profiling?

The last ten years have seen a growing interest in profiling as a means of recording meaningful and informative assessments. In science, there has been a move towards much wider assessment, especially the inclusion of practical skills. If such assessments are to become valid then they must be recorded on profiles.

Profiling is widely spoken about within our schools, and since the Department of Education and Science (DES) has expressed its desire that it should be implemented within all secondary schools there has been frantic activity by many teachers in an attempt to develop such recording systems. In my experience, a stumbling block has often been a lack of understanding as to what is meant by profiling. This chapter undertakes some explanation of the meaning of profiling, by giving the reader an inkling as to how various organisations and authors have attempted to portray it.

Profiling has a diversity of meanings throughout the country. The nearest we can come to an overall definition, therefore, is by looking for common ground in current practices and forming a consensus opinion.

THE UNIONS' VIEW OF PROFILING

The teachers' unions have produced their own publications on profiling.

A National Association of Schoolmasters/Union of Women Teachers (NAS/UWT) publication, *Pupil Profiles* (1982), describes schemes of profiling as being 'designed to reflect a wide range of

1

qualities and skills which have relevance to potential employers'. This organisation sees the use of a profile as a testimonial or reference to be used when seeking employment while this is only seen as one of six uses by the Assistant Masters and Mistresses Association (AMMA).

The other uses on the AMMA list are as:

- references for further and higher education institutions
- reports for the Careers Service, the Manpower Services Commission (MSC) and other external agencies
- an instrument for teachers and parents to monitor pupils' progress
- leaving certificates
- a means of increasing student motivation and encouraging self assessment.

The National Union of Teachers (NUT) sees different advantages and disadvantages for the formative (the continuous assessment and recording of progress) and summative (a record produced once only at, or near, the end of a pupil's compulsory education) profiles. The formative profile is seen by them as an integral part of teaching and learning and a valuable instrument that can lead to appropriate changes in teaching and learning strategies. This type of profile is of little use to employers and is very time consuming for teachers. Summative profiles, on the other hand, are less demanding of teachers' time and can be a useful final record of a pupil's achievements.

The AMMA (1983) sees the summative profile as a much more 'comprehensive description and assessment of a pupil's academic and non-academic achievements, attributes and interests' than does the NAS/UWT, reflecting the much wider use of the profile that is envisaged by AMMA. In *Profiles and Records of Achievement* (1983) AMMA even indicates the nature of the profile's format. It is suggested that achievements, attributes and interests should be 'set out in a format easy to interpret both by educational and non-educational users', and that the profile should be issued 'at the end of the student's period of secondary education'. This organisation further distinguishes between a profile and a record of achievement (as defined by the Government). The latter only lists students' successes and is not intended to be used as a reference, whereas the former is a 'total picture, balancing strengths and weaknesses'.

The NUT publication *Pupil Profiles* (1983) likewise makes it clear that there is a distinction between assessment and profiling, the latter being a 'record of information about a pupil'. Like AMMA it suggests that skills, abilities and achievements, 'as well as other elements which are not formally assessed or assessable', may be included.

OTHER VIEWS ON PROFILING

In the Further Education Unit (FEU) of the DES publication *Profiles* (1982), a profile is regarded as a document that records a wide range of student attributes, such as knowledge, skills and experiences. It is envisaged that common core skills, such as literacy and numeracy, could be tested by conventional means. Furthermore, it is felt that the application of these skills to real-life situations should also be assessed and recorded in profiles.

Worthwhile profiles are seen as those that employ a wide and diverse range of assessment. The approach in this publication is curriculum-led, that is, that profiling should be related to the curriculum and not comprise a series of aptitude and personality tests, which require special skill in interpretation. This is a view that is supported widely by most of those who are actively working in the production of profiles.

The profile is described as a publication for public consumption, designed to be read in its summative form, by employers, parents, education and training personnel and others. In its formative stage it is intended for discussion and reflection between teacher and taught.

The FEU emphasises that profiles should not be represented by a single grade – an attitude that is in tune with current educational trends as will be discussed later. It is important that the profile should be a record that describes the skills, knowledge and experiences of the student in a meaningful accurate way.

In a City and Guilds evaluation report (1983) a quotation from a tutor epitomises the essence of profiling:

> Profiling has developed over the last few years as offering a more equitable way of recording the progress of young people than either public or traditional reports. Profiles seek to describe the progress of young people on a far wider range of criteria than traditional reports, and since they are not bound by subject areas, record things not revealed by public examinations. The criteria of performance, experience and attainment are more relevant to potential employers,

and hence the profile is an appropriate document to pass from one institution to another, or from one institution to an employer.

It is interesting that, yet again, the perceived view of the use of a profile is as a reference, and some of the aforementioned uses are neglected in this statement. This is an indication of the fear of many involved in teaching that unless profiles are accepted as references, particularly by employers, they will not receive wide acceptability and validity. As part of the drive to attain validity for their work some profiling organisations have gone into partnership with other institutions, such as universities and examination boards. One such partnership is the DES-funded pilot scheme of the Dorset Education Authority and the Southern Regional Examinations Board (SREB).

One cannot discuss profiling and profiles without mentioning the influential work of Balogh (1982), which was undertaken for the Schools Council. Here, criteria that yield the most comprehensive and pertinent information for a profile report were set out and surveys were carried out through questionnaires to determine practices in English schools. The criteria that were laid down for a good profile are as follows:

1. The profile report should record the assessment of skills and/or qualities besides traditional attainment, such as assessment of basic mathematical and language skills, cross-curricular skills such as listening and problem-solving, practical skills such as ability to use tools correctly and personal qualities such as punctuality and initiative.
2. This information should be presented in a structured form (though not necessarily graded) and roughly the same kinds of information should be available for each pupil.
3. The profile report should be designed to be given to the pupils when they leave school rather than as a confidential document which is sent direct to users.
4. The profile report should be available to all pupils within the specified target group. School Certificates which were presented to particular pupils who had played a very active part in the life of the school and 'won' them, would be excluded for the purpose of the study.

These criteria have had wide recognition from organisations such as the NUT.

In the surveys carried out in English schools by Balogh 135

schools were contacted of which 75 claimed to have profile schemes in operation, 39 had no profile schemes and 21 did not respond. Over a quarter of the schools that replied appeared to be offering a document more closely resembling a testimonial. Another quarter provided detailed, structured, but confidential reports for the careers service, employers and further education. These were often 'confidential' reports. Others used a scheme based on the Record of Personal Achievement (RPA), which includes no teacher assessment. Only 25 out of the 75 schools were using a profile that approached the criteria outlined above.

The main reason given by 17 schools for developing profile reporting was to recognise the non-examination aspects of school life. Another ten schools wanted to produce profiles to make information more readily available to employers and a further eight wanted to use it as a tool to encourage pupil motivation.

Nine profile report schemes were chosen from the 75 in operation for in-depth study and of these seven met all the criteria described above.

At this point it is important to define precisely some of the terminology that has caused much confusion in the discussion of profiles. In the glossary of *Profiles and Records of Achievement* (Broadfoot, 1986) clear definitions are given for 'assessment', 'profile', 'profiling' and 'records of achievement'. Contributions for the aforementioned glossary were made by the Oxford Certificate of Education and Achievement (OCEA), the National Profiling Network and the DES Records of Achievement National Steering Committee Secretariat. The following definitions are given:

Assessment

An evaluation of a student's achievement. There are many modes of assessment, each designed to allow for the best judgement of a student's performance in a given circumstance. An assessment may be pass/fail or graded or it may consist of verbal reporting.

Profile

A profile is a method of displaying the results of an assessment; it is not a method of assessment. It is essentially derived from a separation of a whole of an assessment into its main parts or components. A profile is a panoramic representation, numerical, graphical or verbal, of how a student appears to assessors across a range of qualities, or in respect of one quality as seen through a range of assessment methods. It is loosely used as a catch-all term for records and reports on pupils' achievements and experiences.

Profiling

A process of deriving information from pupils' experiences and achievements in and out of school, which displays the results in the form of a profile. It may involve negotiation with students and could include a variety of assessments.

Record of Achievement

The term record of achievement is used to describe school-leavers' documents, which may include the results of a variety of examinations, graded tests and other information about a student, as well as internal records compiled by teachers and/or students and covering the total educational progress of the student.

As can be seen from the few examples cited above, there are variations in practices. A more detailed look at the differences and formats will be dealt with in Chapter 4. In the light of the extraordinary range and diversity of profiling schemes that are the result of individual practices throughout the country it is obvious that a universal, standardised scheme may prove difficult to implement. It does not seem possible that teachers would accept a scheme of this nature readily when they have participated in such diverse practices and hold such variable ideologies.

National guidelines for profiling, such as those that are to be drawn up in 1988, can contribute greatly to setting a direction for profiles if they are devised in a spirit of flexibility and some teacher freedom. If guidelines are released as a framework to build on, they will have a greater chance of success than if they are strict unbending rules to follow. Profiling initiatives do, to a large extent, owe their success to the efforts of teachers: the NAS/UWT claim that it is the lack of teacher consultation and participation that explains the limited success of the Scottish scheme. Central guidelines that wipe clean the efforts of teachers will not be well received. On the other hand, having guidelines for profiles will give them wider currency. It is hoped therefore that a dogmatic approach will not be adopted but that account will be taken of such teacher concerns.

A better understanding of the above can only lead to a more successful introduction of profiling. Progress in this area could be greatly hampered by a poor appreciation of what is meant by profiling.

Chapter 2

Assessment

The purpose of this chapter is to look at various aspects of assessment, a word used by Deale (1975) as 'an all-embracing term, covering any of the situations in which some aspect of a pupil's education is, in some sense, measured by the teacher, or another person'.

WHY DO WE ASSESS?

Deale identified the following reasons why the teacher needs to make assessments of his pupils, namely to:

- allocate pupils to sets
- compare progress of pupils under different teachers
- compare new teaching materials with old
- give incentive to learning and an aid to remembering
- inform employers or establishments of higher education about attainment
- inform parents about progress
- decide upon entering pupils for external examinations

It is possible therefore to distinguish three groups of purposes. These are helping the pupil, improving the teaching and providing information for others. In order to assess, a basis for the comparison of performance is required.

THE DIFFERENT BASES FOR COMPARISON OF PERFORMANCE

Commonly, when teachers are assessing, they make comparisons between individuals – this is known as norm-referencing. With this

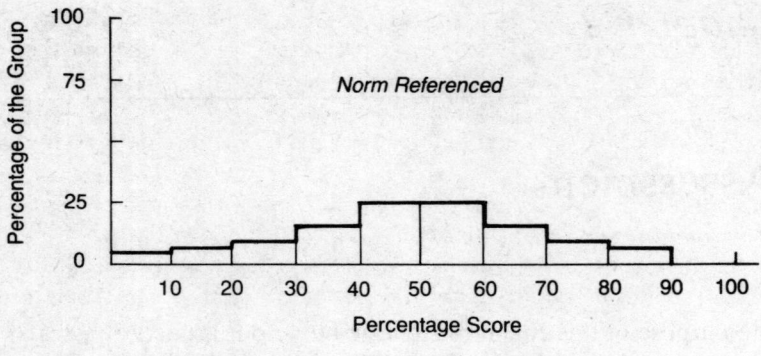

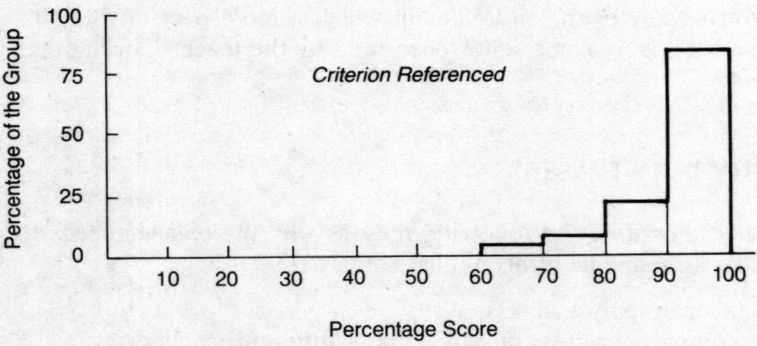

Figure 1 Histograms illustrating the different characteristics of a hypothetical score distribution in norm- and criterion-referenced assessment

type of referencing, the constructor of a discriminating test aims for items that will be answered correctly by about 50 per cent of the pupils. Its major disadvantage is that by differentiating between individuals one is liable to reinforce failure. Also, norm-referencing does not establish, *per se*, any externalised concept of standards.

Comparisons, however, may also be made with external criteria (criterion-referencing). Here, performance of individuals is assessed by comparison with some predetermined criterion. It is also possible to judge students against their own previous performance (ipsative-referencing).

Frith and Macintosh (1984) explain that assessment tends, in practice, to involve elements of both norm- and criterion-referencing: 'A swimming test demands certain levels of performance and is ostensibly criterion-referenced and yet the criteria can change with improvement in levels of performance, for example the qualifying times for the Olympic games are governed by the norm.'

It appears that the dividing line between norm- or criterion-referencing is blurred, and it is not always clear whether any distinction between them arises from the design of a particular test (and therefore reflects purpose) or whether it emerges from an interpretation of test results (which relates to usage).

The norm-referenced test is a discriminating test that aims to discover how much each pupil has benefited from the course. It spreads out the pupils as widely as possible in terms of their ability. The criterion-referenced test is designed to establish how many pupils have achieved a certain standard, or whether an individual pupil has performed a given task. Criterion-referenced assessment has the advantage that the criteria can be pitched at any level. Typically, it is used for guidance and diagnosis, although it is also true to say that the grade boundaries can be adjusted in norm-referencing.

The histograms (figure 1) illustrate the different characteristics of the score distribution in norm- and criterion-referenced assessment.

The norm-referenced assessment has distributed the candidates over the range of marks or grades available while the criterion-referenced assessment has provided a measure of the level of mastery attained by the group. The norm-referenced assessor will be able to tell whether one student is more or less expert than another, but cannot in the end say what aspect of the subject matter has been mastered, that is, he will not be able to say whether this or that objective has been attained.

In order for assessment to take place and to enable the purposes of assessment to be translated into action, a particular approach to the question has to be adopted by the teacher.

APPROACHES TO ASSESSMENT

Two major approaches to assessment can be distinguished. The choice is between the pragmatic and the predetermined approaches, i.e. where assessment is made during the actual teaching–learning situation and where a predetermined plan is used. These approaches are not self-contained and overlap considerably in practice.

The pragmatic and predetermined approaches have been characterised by Frith and Macintosh (1984). The features of the pragmatic approach are:

● assessment of what is actually going on in the classroom
● analysis of the results of the assessment to discriminate between pupils and to ensure that the assessment is well balanced
● the adaptation of techniques of assessment to meet opportunities presented by unexpected outcomes
● the postponement of final grading until all outcomes of assessment can be properly balanced and adjusted.

The predetermined approach, on the other hand, may be characterised by a number of different features:

● the establishment of satisfactory aims
● the setting of objectives by which these aims can be achieved
● the determination of criteria by which the progress or level of mastery of the pupils can be measured
● re-testing and post-testing of assessment material. First, to establish that it is appropriate for the age and the ability range of the pupils. Secondly, to ensure that it is relevant to the subject being taught. Thirdly, to ensure that the results obtained are taken into account
● the determination of ultimate levels for mastery, i.e. pass grades.

The predetermined approach has tended to dominate assessment practices, particularly in large-scale examinations. The stages that are characteristic of this approach are important for the development of any assessment.

The first step in adopting the predetermined approach is to relate a particular course of study to those taking the course. This is done by producing wide-ranging purposes or aims out of which are then selected those measures that are necessary in order to realise the desired outcomes of the course. The broad-ranging and imprecise aims are thus translated into realisable objectives. A classified list of essential objectives is known as a taxonomy.

TAXONOMIES

Bloom classified objectives into three major domains (Child, 1976):

(i) cognitive objectives – placing the greatest emphasis on re-membering, reasoning, concept formation and creative think-ing;

(ii) affective objectives – emphasising emotive qualities expressed in attitudes, interests, values and emotional biases; and

(iii) psychomotor objectives – emphasising muscle and motor skills, and manipulation in all kinds of activities such as hand-writing, speech, physical education, and the like.

The organisation of cognitive objectives will be used to illustrate how the taxonomy operates. The taxonomy (or classification) of cognitive factors was organised by Bloom under six major headings. The six are arranged hierarchically to demonstrate that the objectives are cumulative, so that the higher classes are built on the skills involved in lower classes. The six classes involved are:

(i) Knowledge. This emphasises processes which require recall of specific facts, terminology, conventions and generalisations.

(ii) Comprehension. This represents a level of understanding sufficient to grasp the translation and meaning of mathematic-al or verbal material for the purposes of interpretation or extrapolation.

(iii) Application. This is the use of abstractions and theories in particular and concrete situations.

(iv) Analysis. This means the breaking-down of material into its constituents in order to find the relationships between them. This requires all the previous classes before analysis is possi-ble.

(v) Synthesis. This necessitates the putting together of the consti-tuents by rearranging and combining them so as to give an arrangement not apparent before.

(vi) Evaluation. This involves value judgements about materials, ideas and methods for given purposes. In order to perform this operation satisfactorily, all the skills of knowledge, compre-hension, application, analysis and synthesis are needed.

Taxonomies are frameworks and as such need to be expanded to meet the needs of individual subjects. Later in the text, a taxonomy relating to science will be discussed.

Inevitably, the nature, or what is called the mode of assessment, dictates the educational experience and the quality of the rela-tionship between teacher and students. Assessment acts upon the educational system so as to shape it in accordance with what assessment demands.

MODES OF ASSESSMENT

When the aims and target group for the assessment have been decided, it is valuable to consider a number of modes of, or approaches to, assessment, before the actual techniques are developed. The techniques adopted will differ according to the mode employed. Rowntree (1977) identified eight conflicting modes:

Formal versus informal

The formal mode of assessment has been described as a 'publicly satisfied purpose for public use' and the informal mode as a 'privately specific purpose for private use'.

Formative versus summative

Formative assessment takes place at intervals during the course and is seen as helping the teacher to know the students in order to teach them better. A summative assessment occurs at the end of a course and is seen as a way of helping others to feel better informed about students.

Coursework versus examination

Examination refers to a larger scale test, or, more commonly, a combination of several tests, and perhaps other assessment procedures, whether within the school or conducted by an external examining body. An assessment of coursework may be taken into account as part of such an examination. A student may also be assessed at the end of a course in the light of what has been done during the course.

Continuous versus terminal

Continuous assessment may be used formatively, but may contribute to summative assessment. A teacher could dispense with terminal assessment if a satisfactory summative assessment could be compiled from a series of continuous assessments.

Process versus product

Frith and Macintosh (1984) point out that the ' "product' may fail to reveal all about the "processes" which produced it'. They also state that 'talking about processes is not the same as demonstrating them'.

Internal versus external

Internal assessment (which will be discussed in-depth later in this chapter) is essential if we are to obtain a broad spectrum of information about our students in order to present a profile.

Convergent versus divergent

Convergent assessment is narrow and restricted to the ability of individuals to focus upon a clearly defined task such as is exemplified by the objective test. It is of concern that this type of test will suppress the creative qualities that are encouraged by divergent assessment.

Idiographic versus nomothetic

Idiographic assessment aims to find out about the uniqueness of individuals, whereas nomothetic assessment collects data with the view to comparing individuals. In effect, these terms refer to criterion- and norm-referencing.

Despite the continuing and growing demand for wider patterns of accreditation, of the eight modes of assessment we have just reviewed the examination certificate remains the only documentary record received by most students on the completion of their education. The examination is a very limited form of assessment and the results only reflect certain aspects of the cognitive domain (though they themselves should not be undervalued) at a particular point in time. The mode of assessment should influence the nature of the educational experience, yet we have at the same time an educational system attempting to encourage personal growth and an impersonal system of assessment.

At this point, it is the intention to look at the issues affecting the assessment of science in the light of the previous discussion.

ASSESSMENT OF SCIENCE

In order for students to be able to leave school with documentary records that will present a comprehensive picture of their abilities, interests and experiences, assessment in science should be broadened to include much more internal practical assessment.

Essentially, science is taught through practical methods; pupils are required to use experimental techniques to test hypotheses they have formulated about the behaviour of the materials they are examining and, from the data they obtain, either to reject the hypothesis, modify interest in it, or accept it as valid and build it into theories and generalisations. Carefully controlled 'guided' discovery can allow pupils to establish most of the basic concepts of the sciences by behaving as scientists. In the course of the work they learn many of the processes of scientific thought and action which form the backbone of any good science syllabus. Qualitative relationships, processes, knowledge, the formulation and use of concepts, theories and generalisations, attitudes and practical skills are learned and should somehow be assessed.

Skills required within science can also be found and assessed in other subjects – for example, numerical skills in mathematics; communicative skills in English; manipulative skills in technical subjects; and drawing skills in the creative arts. The list can be extended to cover all subjects, including the social sciences and home economics.

The problem of the assessment of practical skills and the outcomes of practical work is of particular interest, since so little effort has been made at a realistic method of doing this. Deale (1975) suggests that when evaluating an assessment scheme, teachers should ask five key questions:

- Will making this assessment benefit the education of the children, directly or indirectly?
- Is it a valid test of what they have been learning?

- Can it be marked fairly or uniformly?
- Will it provide, when needed, all or part of the appropriate information about the children's attainments?
- Are there any important aspects of the course which are not covered by this assessment?

It has been said by Hodson and Brewster (1985) that 'much of the assessment currently practised in secondary school science departments does not stand up well to this kind of scrutiny'.

The emphasis on the assignment of grades and rankings is regarded as undesirable, since it fails to recognise and reinforce the achievements of the less able. Such gradings and rankings with an over-reliance on norm-referenced testing encourage competition at the expense of co-operation and help to identify large numbers of children as failures.

As already mentioned, testing procedures tend to concentrate on academic, cognitive aspects of the curriculum and the non-cognitive aspects such as individuality, creativity and interpersonal skills are inadequately measured or assessed, so that a balanced picture of a student's strengths and weaknesses is not obtained. The traditional methods of assessment, such as the multiple-choice and structured items tests, tend to concentrate on factual recall and convergent thinking and devalue other skills and abilities that children may possess. Such assessment methods also inhibit curriculum development. Furthermore, it is felt (Rowntree, 1977) that present methods 'often fail to facilitate learning by failing to provide adequate feedback for diagnosis or for the modification of learning experience'.

The present move is to develop assessment methods that distinguish between different kinds of competencies rather than between levels of attainment. Such competencies can be used to produce student profiles where information may bear reference to the knowledge, skills and attitudes of students.

INTERNAL ASSESSMENT

Since the early 1960s secondary school teachers have been directly involved in defining and attempting to solve problems of measuring the attainment of their pupils. Before this time the external examination – the traditional grammar school examination – was the only nationally recognised measure of secondary school attainment. According to the Beloe Committee (Ministry of Education, 1960) this was 'not necessarily appropriate even in the revised form proposed by the Norwood Committee'.

The Beloe Committee concluded that 'despite the dangers, examinations below the General Certificate of Education (GCE) level can make a constructive contribution to the educational process'. When the constitutional arrangements for the setting up of the Certificate of Secondary Education (CSE) Examining Boards were defined, they provided for each region to offer external examinations on syllabuses and examination papers prepared by the Board, and external assessment and validation of examinations conducted internally by individual schools or groups of schools. This shift of emphasis from external examinations to internal assessment and external moderation and validation in the draft constitution of the Examining Boards, is significant.

It is interesting to note that in 1947 the Secondary School Examinations Council had recommended the use of objective tests and systematic internal examinations designed to suit the pupils. In 1962, an inquiry into the nature and purpose of practical work in school science teaching by Kerr, led to the recommendation that the responsibility for the assessment of practical ability should lie more in the hands of the teacher by allowing the schools to set, supervise and mark their own tests.

Schools are now evaluating more and are doing so in relation to what is being taught (Cornwall, 1981). There are probably two discernible sources of influence for this change – the first, the matter of accountability and the second, the growing dissatisfaction with the old model of evaluation along with a related desire for the kind of approach that has greater relevance to the needs of pupils. The new mood at the DES is one that regards assessment as an integral part of good teaching. This is also seen as a good approach to sound management. These two elements – assessment as part of teaching and as an integral activity concerned with management – were closely connected to the general tenor of the DES' views on education starting from the middle of the last decade. The scene was set, therefore, for assessment to become a major issue in education.

Not surprisingly, the DES expressed views that were very close to those of teachers who had for a long time been looking for changes in educational assessment techniques.

The introduction of teacher assessment into science courses provokes a variety of reactions from teachers. A few prefer to stick to the tried and trusted methods of the type of practical work they have long used; more take up the challenge whole-heartedly, recognising that teacher assessment is a logical extension of the educational theories that resulted in such assessment at CSE level, which were reported as being favoured by science teachers and students. Others falter through lack of finance, laboratory equipment or technical help, while still others, although keen, judge themselves too inexperienced to cope with the demands. Probation teachers, new to the classroom, must find internal assessment daunting. Most teachers seem to be agreed on the reasons for and aims of including practical work in their courses. Research enquiries into the views and practices of teaching with regard to practical work in school science have underlined this agreement.

REASONS FOR AND ADVANTAGES OF INTERNAL PRACTICAL ASSESSMENT

In this section, a case will be made for the importance of assessing practical work in science.

Practical work is seen as providing a way, not only of developing a number of different important practical skills, but also of developing favourable attitudes because it provides interest and enjoyment, produces enthusiasm, encourages initiative, imagination and co-operation and develops self-reliance (Sands and Bishop, 1984). It also allows teachers to introduce, develop, extend or reinforce theoretical concepts, to develop scientific method, a critical awareness of experimental design and the ability to interpret data. It enables them to continue to promote the higher intellectual skills involved in scientific thinking when solving problems.

Many teachers place great value on science as a practical subject, commonly regarding this as an important source of pupil motivation. Despite this and despite the fact that science is taught in laboratories and teachers spend a considerable amount of time in supervising practical work, the bulk of science assessment is traditionally non-practical.

In their survey of the context, type and aims of current practice in England, Beatty and Woolnough (1982) found that, for 83 per cent of the teachers in their sample, between 40 and 80 per cent of the time they spent working with 11- to 13-year-old pupils was devoted to practical work. This was seen as an activity that developed practical skills, in particular 'accurate observation and description' and 'deductive logic'. Furthermore, the development of 'problem solving' and 'manipulative skills' and the 'arousal of interest' were identified as important aims. In South Africa, surveys revealed that teachers and educators considered that practical work should serve as a means of facilitating learning and understanding (Lynch, 1978). Despite the 'strong conviction' reported, Lynch noted that, in reality, there was poor implementation of practical work in schools. The National Assessment of Science survey in the USA indicated that laboratory activities were used less frequently than science educators would desire. In Tasmania, Lynch and Ndyetabura (1983) studied the aims of practical work and the order of importance placed on them by teachers: little change took place over the pupil age range 12 to 16-plus years; logical interpretation of observations emerged as the single most important aim at school certificate (15–16 years) and matriculation (16-plus) levels. Moreover, it was suggested that there were some 'major mismatches' between teachers' aims and students' perceptions of those aims. Gunning and Johnstone (1976) reported that Scottish chemistry teachers rated highly, as objectives of practical work, both the pupils' ability to draw conclusions from experimental results and the development of interest in the enjoyment of the subject. These researchers found a 'gulf' between teachers' objectives and the achievement of them by pupils and remarked that 'teachers seemed to have an optimistic expectation of what can be achieved from doing an experiment'.

Teachers' regard for practical science and the major role it plays in our teaching are accurate reflections of its importance. If it is important enough to teach, then we must also assess it. If practical work is not tested, it will undoubtedly be valued less by pupils. There appears to be a mismatch between what we actually teach and what we test. A shift in emphasis has been identified (Bryce and Robertson, 1985) from 1977 to 1982 amongst Australian teachers towards a laboratory-centred approach, but it was considered that there had not been a change in their methods of evaluating students' achievement to accommodate the shift. Obviously, if courses are becoming more practically based, then the assessment techniques must change accordingly to assess the practical skills being taught.

Practical science is important in its own right, in which case, it would seem sensible to assess it directly. On the other hand, it may help pupils to grasp scientific concepts and generally facilitate the achievement of other non-practical objectives. If the latter claim is upheld, it is important that evidence is adduced for it and that there is some understanding of the mechanisms by which practical science has an influence on attitudes and understanding in science. Either way, it is necessary to know what practical science is: once one has described what constitutes it, one might then begin to consider ways in which it might be validly assessed. One cannot simply assume that non-practical assessment is somehow sensitive to what is learned during and from practical study.

Researchers have argued that practical science should be assessed and have questioned the accepted role of laboratory activities. Hofstein and Lunetta (1982) considered the area of practical skills assessment 'a neglected aspect of research' and it is true that, on the whole, most past research studies into the educational effectiveness of laboratory work neglected the important questions: What is the student really doing in the laboratory? What are the appropriate ways to measure a student's activities?

There is a distinction between the development of knowledge and the development of skills. Practical science (as an example of the latter) may easily be devalued if one considers it merely as the medium by which the former is learned. Skills are frequently rendered invisible by our habitual focus on knowledge. Furthermore, there is evidence that those pupils whose achievements are generally poor can do far better at science than the traditional assessment emphasis on recall and explanation permits them to demonstrate (DES/APU, 1984a).

Better descriptions of what happens during practical work are needed, not least in order that we might know how to go about assessing it. Internal assessment should achieve a higher level of reliability, given the greater number of occasions available with this method as opposed to the one-off practical examination. Continuous internal assessment also means that more types of practical activity can be tested, thus giving a better sampling of scientific content and practical skills. In particular, 'handling of apparatus' and 'following instructions' can be assessed, a difficult thing to do when dealing with large numbers of candidates at the end of a course. In addition, students can be assessed on practical work that forms part of a teaching sequence, and the stresses associated with a single examination are reduced or absent, as also are the failure due to accident or chance on a singular occasion. Further, the teacher, when faced with the role of examiner, may well come to consider more carefully his or her teaching technique and any remedial action for individual students.

In spite of the difficulties and problems it was thought that internal assessment would pose, there is a growing opinion among teachers and those concerned with public examinations that the teacher is in the best position to assess his or her own students (Sands and Bishop, 1984). Comparability between the assessment of different teachers can be achieved by the use of moderation – more will be said about this later in the chapter.

As far as teaching techniques are concerned, internal assessment should be seen as a liberating influence. This may seem impossible to the new teacher, but teacher assessment does provide an opportunity for release from the tedious repetition of certain exercises that prepared pupils for the traditional practical examination. Laboratory and field sessions can become more varied, meaningful and relevant to the course as emphasis shifts from training for performance of rehearsed programmes to the development of basic scientific skills.

ASSESSMENT OF LABORATORY PROCEDURES AND OUTCOMES

The principal characteristic that differentiates externally devised and internally controlled practical examinations is the role of the class teacher. In the former, the teacher has little involvement during the examination, though may be required to assess the end-product of an investigation with a mark scheme supplied by the examination board. In the latter, the teacher is involved in the design of the experiment and the observation and assessment of practical skills that may be recorded in various ways.

In one type of examination, marking can be done according to a predetermined key of weighted scores for skills of manipulation, self-reliance, communication, experimental design and measuring. At several points during the examination, the examiner would mark students' answers and provide continuation cue cards. Another type of practical examination uses the 'stations' technique (pupils circulating round a number of small 'experiments' or practical tasks usually on a timed basis).

Internal assessment need not be restricted to terminal examinations and indeed most would argue that it should not be so constructed. It is believed by many teachers that greater gains may be achieved if the assessment is continuous and integral to the teaching of practical science, in which case the assessment is formative or diagnostic in character. To be genuinely continuous, the reporting procedures must not penalise early efforts, where practice has still to take place.

Up to now, practical examinations have been based on written reports and continuous assessment has often been interpreted as the grading throughout a course of written reports based on practical work. Some educators have argued that what is needed is a clarification of the aims and objectives of the practical work with a programme of investigations designed to allow the objectives to be achieved. Here, laboratory reports may be valid in conjunction with the direct observation of students by the teacher.

Authors such as Kelly (1971) warned against wrong assumptions about skill acquisition from reports of results. There tends to be an overemphasis on neatness, writing skills and the volume of written reports. Reports, moreover, cannot provide direct information on skills in manipulating equipment, observing or organising and performing an investigation efficiently. Despite these reservations, Sands and Bishop (1984) believe that one ability might be assessed during a practical session using checklists (these will be illustrated later) with, or replaced by, impression grading while another ability might be assessed afterwards from written reports or oral questioning. They envisage the practical work that is to be assessed being done as part of the teaching of a particular topic.

Besides normal laboratory practical work there are other types of activity that can be used. A teacher demonstration, particularly where the practical involves the use of large or expensive apparatus, can be structured so that students in the class make their own observations and interpretations. An extension of this technique is to use data that have not been gathered at first hand by the class. A-level science students undertaking Nuffield courses are given wide experience in such methods through the Study Guide work.

Project work and field work are also used for assessment, but the fact that they are long pieces of work does not accord them a greater weighting than other practicals.

The mechanics of the various schemes of internal assessment in science courses are provided by individual examining boards and techniques of procedure and grading vary from one board to another. However, some general points can be made and are worth emphasising.

Lots of practical abilities given for assessments by examining boards show that some are procedures and some are outcomes, that is, process and product skills. The process skills in laboratory activities usually have to be observed in order to be assessed, and are probably assessable only by the teacher. The products have, in the past, formed the basis of practical assessment made as a result of external practical examination where it was assumed that the process by which the product was reached was the skill ostensibly being tested. The procedures include such skills as manual dexterity and manipulative skills; the handling of equipment, materials and organisms; the use of scientific techniques; the use of instruments; the ability to follow instructions, understanding and adapting as necessary; observation and identification; the taking and clear ordering of results; the practical procedures that go with experimental design, and so on. Even oral contributions to group discussions of a practical problem can be included here.

Such procedures, which are so difficult to assess during the course of a traditional practical examination, become possible with school-based teacher assessment. The activities must be assessed during the practical as the teacher observes the students at work and notes their progress. A schedule or checklist of points may be used; this is now common practice for many science courses. Checklists give systematic, directed observation that is a great deal more useful in assessment than random, undirected observation. Schedules used in the individual practical tests administered as part of the work of the APU (to be discussed more fully later) showed very clearly the precision and exactness possible in the careful construction of such lists when scientific behaviours are being observed. Checklists also enable a teacher to pinpoint more readily student weaknesses, ineffective instructions or problems with the worksheet.

For practical skills that seem to have a broader basis, a closely defined checklist is of less use. Manual dexterity, for example, involving the rapid and confident completion of tasks, or the ability to work in a methodical fashion involving correct sequencing of tasks, effective use of materials, good use of time and the ability to adapt instructions to one's own way of working, do not readily lend themselves to a detailed checklist. In such circumstances, an impression mark on a 5- or 10-point scale (or A–E) is usually given.

The products of practical work may be presented in a number of different ways including diagrams, tables, graphs, calculations, statistical analyses, written accounts, answers to questions, as well as practical end points such as a finished dissection or a chromatogram. Here checklists of points can be prepared in advance and applied to the material being assessed.

In addition to the usual list of skills that are assessed in science, assessment of attitudes, that is, interest and enthusiasm, persistence in trying to resolve difficulties, resourcefulness, co-operation in normal laboratory routine, and respect for living organisms (especially applicable to biology) and the environment may also be assessed.

It is appropriate at this point to look at a specific example of an 'old-style' practical examination paper and discuss what is being assessed. The exemplar (see Appendix 1) is a 'traditional' end-of-course examination paper, set by the Cambridge Examinations Syndicate for the GCE O-level biology examination for November/December 1982. This is one out of four examination papers and is marked by an external examiner who awards a mark out of 40. Students were given one hour to answer all the questions.

In order to establish the skills that were being assessed by this paper, the checklist produced by the Inner London Education Authority's (ILEA) Further and Higher Education (FHE) Curriculum Development Project (see Chapter 4) is used here. Out of the 23 skills listed on the checklist, only the following seven were being assessed:

P1   using observation
P4   measuring in appropriate units
P5   calculating using appropriate units
P6   drawing and using representational diagrams, charts, symbols and formulae
P9   suggesting explanations, causes or solutions
P13 recording observations and procedures in suitable form
P14 drawing conclusions at an appropriate stage.

Even though ILEA's Development Project calls all the skills on the checklist process skills, some are outcomes of practical work. In fact, all of the above categories being assessed are outcomes (or products) of practical work. None of the affective or process skills were tested for.

There are other problems associated with such a paper. Minor accidents or mistakes can cost the student dearly, since these will affect the results of the practical and most of the marks rely on obtaining the correct end-product. Also the food tests required on this paper can unwittingly be an area of error for many students. In the specific example given, a piece of apple was tested for reducing sugar. The result obtained from such a test can vary depending on the variety of apple, the age of the particular fruit or other individualities, such as storing techniques. An examiner's mark scheme often deals inadequately with such variation.

The same criteria can be applied to many other 'traditional' practical examinations and similar findings will be made, to a greater or lesser extent.

We now turn to the major issues affecting practical science. Each issue will be discussed in turn.

MAJOR ISSUES IN PRACTICAL SCIENCE ASSESSMENT

Norm- and criterion-referencing in the assessment of practical science

Some general discussion of norm- and criterion-referencing has already been made. This section is concerned with assessment in practical science.

Those assessors who use scales (1–5 or A–E) with no further grade descriptors effectively concentrate their efforts on discriminating between the performance of students. Such devices are norm-referenced. Among other problems associated with norm-referencing (as previously mentioned) it is well known (Bryce and Robertson, 1985) that teachers avoid awarding the extremes of any rating scale (but they are often encouraged by the examination boards to do so). When teachers feel this compulsion to use the whole scale, discrimination between students is achieved, irrespective of the performance of individuals. It may be that in reality a group deserves grades ranging from A to C, yet in order to use the scale fully, some teachers will award an E to the low Cs and 'stretch out' the distribution of grades accordingly. This is an unfair and an unacceptable state of affairs. With grades that are defined by criteria a more realistic identification of what students can do will be possible. Criterion-referencing enables the teacher to focus upon the performance of students against a list of skill descriptors.

With a gradual shift from external to internal practical assessment we have an associated shift from grading (rating-scales) to the use of checklists (or at least grades with descriptors). This move is both part of the increasing recognition that criterion-referencing is desirable if assessment is to be reliable and the acceptance that such assessment has favourable effects on teaching.

In impression grading of practical work, such as that in the study by Kelly and Lister (1969), only outline guidance was given to teachers on how to carry out the grading of three skills – laboratory procedures, recording and handling results. Nevertheless, they claimed that their overall assessment procedure was comparable in accuracy to a series of practical tests. This may be true for experienced teachers, but would those new to teaching achieve such admirable results? Details of the Nuffield A-level biology scheme may be seen in Appendix 2.

Bryce and Robertson (1985) refer to the advice given to teachers on how to carry out assessment as stated by the Schools Council (1973) in preparation for a GCE trial:

> The following will be assessed:
> (a) Manipulative skills
> (b) Skills in observation and accurate recording of observations etc . . .
> It is intended that teachers should choose their own method of making assessments which, it is hoped, can be made as unobtrusively as possible during their own teaching.
> Each ability should be assessed on a five-point scale according to the following scheme:
> 5. Outstanding   4. Good   3. Average   2. Weak
> 1. Poor (Average refers to the average A-level candidate)

Eglen and Kempa (1974) stated that assessment procedures 'based largely on impression grading', lead to considerable differences in grades awarded by teachers for the same practical performances due to the differing criteria (with major differences in weighting) adopted by the assessors. It was felt that checklist schedules would help to standardise procedures. It has also been felt that during their assessments teachers are giving all-round judgement that reflects the results of the written examinations.

As I am a biology teacher, examples from biological work will be used in this section to illustrate certain trends in assessment. In 1979, the Associated Examining Board (AEB) and the Joint Matriculation Board (JMB) provided short descriptions for some or all of the scale points applicable to one or several of the abilities to be assessed. The AEB identified four areas of practical ability to be assessed in A-level biology and attempted to be more specific about what constituted each area, although instructions to the class teacher on how to carry out the grading remained vague. The specifications were as follows:

A. The ability to apply experimental techniques.
B. The ability to observe and record results of investigations.
C. The ability to interpret observations and draw conclusions.
D. The ability to plan and carry out investigations.

Criteria for each ability were established. To test each of the four abilities the following scale was applied:

5 marks   –   Has reached independence in effectively demonstrating competence.
4 marks   –   Can carry out most of the listed procedures effectively and with minimal assistance from the teacher.
3 marks   –   Can successfully carry out a number of procedures but only with considerable assistance from the teacher.
2 marks   –   Finds difficulty in achieving any significant success, even though an attempt is made with considerable assistance from the teacher.
1 mark   –   Makes little attempt to undertake work or to follow through a series of activities.
0 marks   –   No acceptable work produced.

It was recommended that each ability was tested at least twice, the average for each ability to be then totalled and doubled to provide a mark out of 40.

Further description of each ability of the updated scheme (1987), may be seen in Appendix 3.

In the recently implemented (September 1986) General Certificate of Secondary Education (GCSE) syllabuses, similar approaches have been developed. One such example is the Midland Examining Group (MEG) biology syllabus. Here six skill areas are identified:

1. Following instructions
2. Handling apparatus and materials
3. Observing and measuring
4. Recording and communicating
5. Interpreting data
6. Experimenting design/Problem solving.
   For each skill, the marks awarded are allocated thus:

$$H\begin{cases} 9 & \text{high} \\ 8 & \text{intermediate} \\ 7 & \text{low} \end{cases}$$

$$I \begin{cases} 6 & \text{high} \\ 5 & \text{intermediate} \\ 4 & \text{low} \end{cases}$$

$$L \begin{cases} 3 & \text{high} \\ 2 & \text{intermediate} \\ 1 & \text{low} \end{cases}$$

H, I and L stand for high, intermediate and low respectively. Marks are awarded for each skill area according to specified criteria. These criteria and fuller details on the methodology may be seen in Appendix 4. This scheme will be revised with a common scheme for all sciences to take effect from 1990.

As can be seen, grade descriptors have assumed considerable significance nationally as is indicated by the efforts of examination boards to change the basis of their certification procedures. Bryce and Robertson (1985) point out that a

> shift in the basis of certification (in the direction of criterion-referencing) would seem to be possible if GRC [grade-related criteria] statements are truly descriptive; that is, if they state what pupils should be able to do (without reference to the performance of others).

The authors further explain the difficulties involved. They suggest that capabilities are 'many and varied and impossible to circumscribe in a "few" words' and that such capabilities are 'rarely, if ever, conceived in descriptively discrete steps which might match the "graduations" of a grade level system'. They conclude that in the case of practical science skills, 'the GRC "model" to be workable only if the detail embodied is akin to that obtained with a conventional objectives model'. Here, Bryce and Robertson are referring to the setting of precise criteria for each grade, in the same way that objectives are determined for courses. Perhaps the difficulties that have been experienced in the setting of criteria for GCSE is an indicator that this is going to be no easy task.

Moderation

As standards will inevitably vary among teachers, it is essential to relate the teacher-assessed mark to some common standard and the success of a scheme involving internal assessment depends on moderation. Moderation is the process by which comparisons can be made between different groups of students assessed by different teachers in different schools with varying facilities. In many instances, teachers use their own set of laboratory practicals and assess in their own individual way. In the process of moderation the standards and marks awarded by individual teachers are equated with the whole student population being assessed.

Methods of moderating practical work

● A series of standard mini-tests of prescribed exercises.
Such mini-tests could be provided by the examining board together with detailed objective mark schemes. One such scheme was introduced as an option to a practical examination by the Cambridge Examinations Syndicate in September, 1985 (see Appendix 5). Teachers were expected to implement these tests and follow the scheme of marking to give a set of marks that would hopefully be equivalent for all candidates. This eroded the freedom and flexibility of the teacher's internal assessment; assessment became formal and externally set.
● Moderation by inspection.
The work of all, or of a sample, of the students is inspected in order to compare it with others, by a moderator either re-marking the work or visiting the school. A visiting moderator makes personal judgements about the candidates' work and compares it with a 'mental standard'. The system of inspecting relies on the experience, judgement and integrity of the moderators, and on their ability to reliably relate students' work to some notion of suitable standards.

Although this is a useful method, it is not practicable on a large scale, due to the problems of organisation and cost.
● A statistical method of moderation.
Where large numbers of students are involved, a statistical method is the most widely used. A moderating instrument, such as a practical test (unlikely) or a theory paper, which is an examination component taken by all candidates, is used. Usually, the students' performance in the examination designated as the moderating instrument is used to scale the teacher's marks for the same students in terms of standard and range of marks. The school's rank order is not changed; any alteration to the teacher's marks results in a moving up or down of every student's mark.

Written papers that are used as moderating instruments could test a sub-set of the skills assessed by the teacher and other skills that are associated with the laboratory work and that form part of the teacher's assessment, such as data analysis and interpretation, hypothesis formation and the planning of investigations.

Evidence (Gunning, 1978) casts considerable doubt on the validity of the practical assessment. It is claimed that there is a real possibility that what is being assessed by teachers is not what is intended.

REASONS FOR THE LACK OF SUCCESS OF PRACTICAL ASSESSMENTS

The lack of an overall conceptual framework

It is necessary to understand what is meant by practical science in order to be able to assess it. Doran (1978) identified the lack of 'organisers' for the skills and abilities described as a limitation to the success of the behavioural objectives approach. Various organisers have been suggested; Jeffrey (1967) identified six areas:

    communication
    observation
    investigation
    reporting
    manipulation
    discipline

Other alternative categories have been suggested such as: measuring, identifying, selecting and computing. Kempa and Ward (1975) identified the planning and design of an investigation, the carrying out of experiments, observations of particular phenomena and changes and the analysis of results.

The 'practical mode' has been described by others as three distinct modes: problem solving ability, skills in the performance of routine laboratory tasks and the ability to make observations. Such categorisation is essential in order to focus the assessment on particular components of practical work. It is interesting to note (Eglen and Kempa, 1974) that: 'correlation between proficiency with which a practical task is performed and the quality of results derived from it is low'.

The lack of skill definition

Another problem is the actual definition of the skills. Sometimes some attempt at defining the skills has been made, for example the AEB (see Appendix 3) and the JMB (1978) describe the skills in their scheme for internal assessment of practical abilities in A-level biology. This is the JMB's attempt to organise the practical skills (reproduced with permission):

The JMB scheme for internal assessment of practical abilities
Each of the four abilities (A, B, C and D) is to be assessed on a ten-point scale (1 to 10).

A.   *Possession of appropriate manipulative skills*
There are four skills to be assessed under this heading. It is not intended that these skills should necessarily be assessed as separate exercises.
(i) The use of dissection instruments.
(ii) Cutting sections, mounting and staining of temporary preparations.
(iii) The use of a lens and a microscope (both low and high magnification).
(iv) The handling of apparatus.

B.   *Carrying out observational investigations*
Under this heading the student is to be assessed on the ability to exercise powers of investigation, to identify, to interpret and record important features both microscopically and macroscopically and, where appropriate, to make comparisons. The skills listed under A will be involved in the investigations, and care must be taken to avoid confusing assessments of the two categories.

C.   *Carrying out experimental investigations*
Under this heading the student is to be assessed on:
(i)   The ability to carry out investigations in accordance with specified procedures.
(ii)   The presentation and handling of results.
(iii)   The interpretation of results.
In handling data, the student should be required to draw quantitative conclusions from the data by making simple calculations. In handling data, it is preferable that the student should use results obtained within the class as often as possible. In interpreting class results, the student must keep in mind the actual situations and constraints that this imposes on what conclusions might be drawn.

D.   *Planning investigations*
The assessment should be based on the student's ability to plan an investigation with proper regard for the limitations of the methods proposed, the need for controls, the choice of a form of

presentation for the results, and the weight to be attached to such results.

This ability is not intended to be interpreted as the extent to which a student can satisfactorily design and carry out an independent research project.

Sands and Bishop (1984) developed a hierarchic taxonomy of the objectives of practical work with suggestions of methods by which assessment could be undertaken. In their model the objectives are cumulative so that higher classes are built on the skills in the lower classes. The fundamental difference between this approach and the previous one is that the skills here do not 'stand' on their own. Students are not assessed in the higher skills unless they have been successful in the preceding ones. For example a student will not be assessed on the 'knowledge of procedures' category unless he can name the apparatus. It is important in assessing students that we remember that it may be of little value and unfair to the students if we try to assess the higher levels unless they have mastered the lower ones.

The following is the hierarchic taxonomy developed by Sands and Bishop (1984) (reproduced with permission):

A. Knowledge of apparatus

All that would be required under this category would be the ability to name practical apparatus and to state its purpose in terms of use. Assessment could be undertaken by presenting the pupil with pieces of apparatus and requiring him or her to name and to describe their use.

B. Knowledge of procedures

All that would be required here is basic knowledge of routine procedures, i.e. the extent to which the pupil knows the procedures for carrying out routine practical operations which are basic to the subject. Assessment could be undertaken by using questions requiring descriptions of the procedures.

C. Knowledge of ways of using apparatus

Under this heading the pupil would be expected to know how to use apparatus involved in carrying out routine procedures and how to handle the apparatus to achieve varying degrees of accuracy. Knowledge and use of safety precautions would also be included.

D. The ability to use apparatus

Under this heading the pupil would be required to show that he could use apparatus of which he had knowledge. This means that he should be able to combine the aspects of knowledge required under A, B and C and to apply some degree of manipulative dexterity in the performance of relevant operations. Assessment for these last two categories could be undertaken by means of simple exercises involving the use of the particular piece of apparatus. The criteria for judging the quality could be based either on pre-determined results with an allowed margin of error and/or direct observation of the work being carried out.

E.   The ability to implement procedures
This heading would involve the carrying out of those procedures of which the pupil had knowledge (B).
Assessment could be undertaken by providing simple instructions for the particular procedure and requiring the pupil to perform it. The assessment would take into account the extent to which the pupil carried out any checks necessary to ensure the satisfactory working of the apparatus used and the necessary accuracy of results.

F.   The ability to select appropriate procedures for a particular problem.
Under all the previous headings the apparatus or procedures have been in the limited context of a specific of specified purpose. This heading involves the ability of the pupil to select, by applying the appropriate criteria, the most suitable apparatus and/or procedure for a particular practical or experimental task.
Assessment could be undertaken by presenting the pupil with a practical problem and a range of alternative apparatus and procedures. The extent to which most appropriate choices are made would form the basis of the assessment.

G.   The ability to observe the material under investigation
Under this one heading there are two levels of complexity. At its simplest level the ability is concerned with the identification and classification into known categories of the objects or processes which are being investigated, i.e. a descriptive skill based on observation. At a more complex level the ability involves the pupil in the exercise of a degree of discernment in establishing patterns from his observations and systematising the information derived from them.
Assessment could be undertaken by requiring the pupil to translate his observation into oral or written terms. The criteria adopted for the measurement of the ability would be the extent to which the pupil's observations were comprehensive, systematic and ordered in terms of significance.

H.   The ability to observe changes or differences taking place in the material under investigation
This heading is an extension of G. There, attention was to be paid to the general method and effectiveness of the pupil's observation; under this heading is included the extent to which the pupil is able to recognise the changes that take place in the material being studied and, having recognised such changes, to take whatever steps are necessary to examine them systematically.
Assessment could be undertaken by presenting the pupil with a practical situation and requiring him to observe the changes

which take place and to identify and isolate the changing factors in such a way that they can be studied more comprehensively, e.g. by refined counting techniques and sampling procedures and by eliminating possible errors and variations resulting from the techniques used or from inherent variation in the material being studied.

I.   The ability to record appropriately observed material and the changes which take place in it.

This heading covers the ability of the pupil to make and keep records of the activities described in G and H.

Assessment could be undertaken by evaluating the use made by the pupil of the available methods of recording observations. The criteria upon which the assessment is made should include not only the pupil's selection of the available methods of recording but also the extent to which the pupil is aware of the possible distortions of the data by the use of different methods of presentation. They should also include the extent to which conclusions drawn from the data arise from the data itself or are conditioned by the method of presentation.

J.   The ability to devise new apparatus or techniques to meet the demands of a particular problem

This heading involves the ability of the pupil, when faced with a practical problem which cannot be solved satisfactorily by the use of familiar apparatus and procedures, to make modifications and adaptations to known apparatus and procedures to meet the demands of the new problem.

Assessment here requires the pupil be presented with a situation in which the apparatus and techniques used and the problem posed are sufficiently within his experience for him to appreciate that the former are insufficient to solve the latter.

K.   The ability to plan and carry out a practical investigation

This heading involves all the preceding categories and the use of all the pupil's practical experience and skill in the design of practical work and its execution.

Assessment can be undertaken by presenting the pupil with a problem and requiring him to plan an appropriate practical procedure to solve it. The initial plan should be written out by the pupil and evaluated before it is implemented. The assessment should take into consideration the extent to which the pupil has anticipated all the problems which he could justifiably be expected to anticipate at the planning stage. It should also include the extent to which the pupil recognises problems as they arise and modifies his plan to overcome them. It is therefore not necessary to ensure that the plan is foolproof before the exercise begins, but care must be taken to ensure that the plan which has serious faults when first designed is modified to enable the pupil to proceed with the practical work.

L.   Attitudes to practical work

In almost all subjects in which it is felt necessary to incorporate practical work in the overall pattern of assessment, one of the most important educational objectives is the establishment of particular attitudes towards practical work. The desirable attitudes here are, for example, 'willingness to co-operate in the normal routine of a laboratory', 'persistence', 'resourcefulness', 'enthusiasm', 'the ability to work as a member of a group', 'commitment to practical work as a worthwhile pursuit without compulsion'.

Attempts to assess such qualities are rarely included in public examinations because of the very real problem of making assessments with any degree of objectivity. These qualities, moreover, are those upon which it is particularly difficult for individuals to reach agreement over their recognition and definition and upon the standards to be applied in assessment if they can be identified. It is, however, necessary to note that this area of attitudes is probably the most important of all in terms of teaching objectives. Careful consideration should, therefore, be given to the matter before a decision is taken to include or exclude what in Bloom's terminology are 'affective' objectives.

It is interesting that Sands and Bishop include the assessment of attitudes to practical work, something which many have avoided because of the problems of assessing affective skills. Regarding attitudes, Hodson and Brewster (1985) draw an important distinction between 'attitude to science and scientists and the scientific attitude', which is defined as 'an attitude to ideas and information and to particular ways of evaluating them'. If we are to assess in order to produce profiles, then many teachers believe that we need to take account of this distinction and that despite the difficulties involved in the assessment of attitudes, they are skills (known as affective skills) that need to be assessed if we are to build up useful profiles on individual students. In recent years, there has been an increasing trend for examination boards to introduce syllabuses that attempt to assess attitudes (see Appendix 5).

The lack of strategies for implementation of assessment schemes

In the past, examination boards did not give clear indications as to how assessment schemes would be implemented by the classroom teacher. The boards claimed that they were giving the teacher the freedom to devise practical work to suit their individual inclinations and circumstances. It is now felt that such boards had no clear idea of how their assessment scheme could be implemented in a real classroom/laboratory situation. This is exemplified by a recent occurrence during a Phase 3 GCSE training session, in which science teachers were told, 'You are not here to be told how to put GCSE assessment into practice; that is for individual teachers and schools to work out'. On this question of providing detailed information, Brown (1980) points out that to merely inform teachers that diagnostic assessment can be informally carried out is inadequate and that advice on the ways that it can be done is needed.

Lack of official guidance has meant that teachers have adopted different styles and methods of assessment, making assessment subjective and unreliable. Gunning (1978) reports that teachers complained that much of the coursework did not lend itself easily to assessment; topics were found to be unsuitable for assessment or experiments were too short and undemanding. It was noted that fewer assessments were made by teachers than had been anticipated.

The idea of a planned, integrated teaching/assessment situation is not new, but in the past, most of the assessment procedures used were tacked onto existing and usually heavily content-laden courses. Teachers expected, for this and many other reasons, a reduced content and more relevant courses when GCSE syllabuses were introduced – many were disappointed.

Nevertheless, the 'new' assessment approach has had its impact. This impact may be seen in new courses such as the Certificate of Pre-Vocational Education (CPVE) the *Consultative Document* of which calls for both formative assessments, which provide evidence of young people's attainments to indicate future learning needs, and for summative procedures, which measure those attainments that the curriculum seeks to develop within the contexts in which they are learned (CPVE, 1984).

WHAT DEVELOPMENTS ARE NOW TAKING PLACE IN THE ASSESSMENT OF PRACTICAL SCIENCE?

These developments are only discussed in outline here, since the major points are dealt with elsewhere.

Production of detailed checklists

Some of the merits of detailed checklists have already been mentioned. Such checklists should enable a teacher to make valid assessment of certain practical skills; it can be difficult, though, to operate a checklist for each member of a class simultaneously. The answer to this problem may be to assess only a small group of individuals at any one time. With experimental, continuous work, furthermore, checklists can be difficult to put into practice because while a number of checklists are being operated pupils often work together. Any procedure must offer the possibility of an objective and valid assessment of each individual pupil within the class. Tests such as those produced by the APU may serve to provide the answer to this problem.

Production of product checks

In the Scottish initiative, *Techniques for the Assessment of Practical Skills in Foundation Science* (TAPS) (Bryce et al., 1983), sequential sets of highly structured tasks have been developed. The use of checklists for which an observer was required were limited to a small number of important basic laboratory procedures such as using a bunsen burner or thermometer. Product-based 'end check' test items have been developed to indicate ability on the bulk of the course objectives. The researchers believe that for many teachers, the attractions of a manageable system (which 'stations' constitute) have been the decisive factor in the acceptance of practical assessment. The practical test items, moreover, are seen as a worthwhile first strategy for teachers in gaining familiarity with pupils' strengths and weaknesses after which a more integrated approach of assessment can then be adopted.

It is obvious that much more developmental work is needed in order for teachers to integrate their practical assessments more closely into their teaching.

Assessment for teaching

It is essential that teachers should be able to develop techniques for practical science assessment: 'Assessment is for teachers' purposes and the needs of examination boards should be subservient to them' (Bryce and Robertson, 1985). It is important in this development that well thought out grade related criteria are established and that there is the back-up and material support for teachers carrying out internal assessment.

Internal assessment has become an international preoccupation. Chapter 3 will focus upon three British schemes that have produced methods for the assessment of practical skills. The chapter will indicate how some of the previously mentioned assessment techniques have been implemented. Practical guidance to the use and production of checklists for GCSE assessment will also be given.

Chapter 3

Assessment In Practice

In this chapter the methods of assessing practical skills as developed by three British schemes of national importance are highlighted. The inherent methods of recording are also illustrated. This is intended to be a practical chapter, of particular use for teachers concerned with GCSE assessment. Guidance will be given on how the teachers can attempt GCSE assessments in the easiest, most practicable and worthwhile way.

AN OUTLINE OF NATIONALLY IMPORTANT ASSESSMENT INITIATIVES

The Assessment of Performance Unit

The Assessment of Performance Unit of the Department of Education and Science was set up in 1975 to monitor performance in several of the major curricular areas. The unit has stressed that it is concerned with a cross-curricular picture of national performance, that is, the achievement tests will be related not to specific curricula but to the general objectives of the curriculum. The areas that are the subject of tests are English language, mathematics and science. Other areas in which monitoring is in existence are foreign languages and design and technology. In science, the tests are administered to national samples of 11-, 13- and 15-year-olds.

The APU science project teams began work in 1977. In the time leading up to the first national surveys, in 1980, the project teams and steering group devised and refined, through trial and consulta-

Table 1 *The categories of science performance*

Category	Sub-Categories	Form of test
1. Use of graphical and symbolic representation	– reading information from graphs, tables and charts – representing information as graphs, tables and charts	Written test
2. Use of apparatus and measuring instruments	– using measuring instruments – estimating physical qualities – following instructions for practical work	*group practical test
3. Observations	– making and interpreting observations	group practical test
4. Interpretation and application	– (i) interpreting presented information – **(ii) applying: Biology concepts Physics concepts Chemistry concepts	written test
5. Planning of investigations	– planning parts of investigations – planning entire investigations	written test
6. Performance of investigations	– performing entire investigations	individual practical test

Source: DES/APU, 1985a (reproduced with permission).
*Category 2: At age 11, Use of apparatus and measuring instruments is assessed in the form of an individual practical test.
**Category 4–11: At age 11, Application of concepts in physics, chemistry and biology are not tested separately but as applying science concepts to make sense of new information.

tion, the assessment scheme. The scheme comprises a set of six main categories and sub-categories, describing scientific processes of enquiry, such as 'observation', 'interpretation of data', 'planning investigations', 'performing investigations', 'use of graphical and symbolic representation' and 'use of apparatus and measuring instruments'. In defining the performance to be assessed, a list was drawn up of the concepts and knowledge used in selecting content for some of the test questions, particularly those in the sub-category 'application'. Three of the six main categories are assessed in the practical mode. (See table 1.) The form of test, referred to as a 'group practical test', is in fact a circus of activities where pupils work individually.

Process skills, concepts and attitudes are regarded as important areas in the development of children at the age of 11, 13 and 15 and therefore need to be included in the assessment programme. These skills are those which 'provide children with ways of finding out about their world' (DES/APU, 1984b). This is done by 'seeking and using evidence by observation or investigation, by interpreting information, drawing conclusions and applying ideas to new problems'.

Students may gain knowledge through the application of skills and with the help of existing knowledge this may help them make sense of what they observe. Concepts may develop from the relationship between different ideas. The APU sees scientific development as a 'gradual building of a framework of ideas which are used in making sense of further experience' (DES/APU, 1984b). Furthermore, it espouses the idea that attitudes such as open-mindedness, willingness to take account of evidence, to be persevering and to be critical in thinking are needed if pupils are to maximise the use of their skills and concepts. This, though, represents a particular view of attitudes that requires substantiation by further research.

The APU (DES/APU, 1984b) states that:

> by the age of 10 to 11 years, children require not only the basic concepts widely applicable to all activities (e.g. length, area and time) but also the more specifically scientific ones (about living things, movement, forces,materials, etc.).

In order to be able to say what children can do well and what they do less well, the skills have to be assessed separately as well as in combination during the investigation. The APU have done this by defining categories of performance (see table 1) and developing test questions relating as closely as possible to one category only.

The categories used by the APU exemplified

Category 1 (figure 2) use of symbolic representation.
The ability to put information in symbolic form and also to interpret information written in symbolic form are important in scientific development.

Sub-category   reading information from graphs, tables and charts. Example:

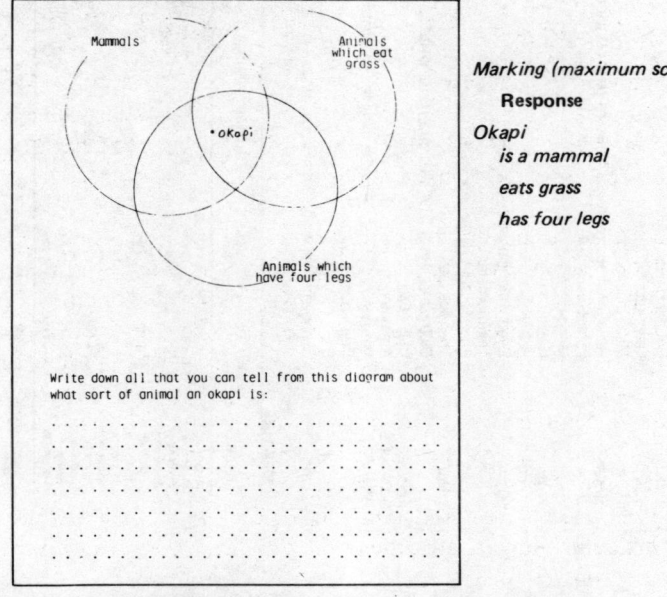

Write down all that you can tell from this diagram about what sort of animal an okapi is:

Marking (maximum score = 3)

Response	Mark
Okapi	
is a mammal	1
eats grass	1
has four legs	1

Figure 2   Reading information from a Venn diagram

Source: DES/APU, 1984b. (Reproduced with permission.)

Sub-category   representing information as graphs, tables and charts.

Category 2 (figure 3) use of apparatus and measuring instruments. In some circumstances, this category may be a prerequisite for other skills to be demonstrated such as making observations and carrying out investigations.

Sub-category   Using measuring instruments.

Using measuring instruments

Ten different measuring instruments have been set up all ready for you to look at.

Do not change them in any way.

Answer as many of the questions below as you can.

Don't forget to say what units you are using.

This is an example answer to show what we mean.

How wide is this page?   .... 21 cm ....

a) How much water is there in the measuring cylinder?   ....

b) How big is the force with which the rubber bands pull on the hook?   ....

c) What is the mass of the object **X** ?   ....

d) What is the length of the pencil?   ....

e) What is the temperature of the water in the flask?   ....

f) How long had the stop-clock been running before it was stopped? (It started at zero.)   ....

g) Press the push-button to switch on the current. Read the current through the circuit on the ammeter.   ....

h) Press the push-button again. Read the voltmeter.   ....

i) How much liquid is there in the syringe?   ....

j) What is the pressure of the gas at this gas point?   ....

APPARATUS

1	100 cm³	Measuring cylinder in cm³
1	50 N x 2 N	Forcemeter
1		Retort stand
2		Bosses
4	thick	Rubber bands
1	0 - 200 g	Lever arm balance
1	118 g	Mass labelled 'X'
1	30 cm	Ruler, marked in mm, cm. Sawn off pencil stuck along top from 3 cm to 9.4 cm
1		Manometer
1	1 m	Gas tubing
1	50 cm³	Bottle containing
1	50 cm³	Dyed water (fluorescein solution)
2	-10 to 110°C	Thermometer in
2		Bung bored to fit thermometer
	½ litre	Thermos flask
		Stop-clock, set at 7'17" with buttons removed
2	4.5 V	Batteries
	0-1 A x 0.01	Ammeter
	0-5 V x 0.2	Voltmeter circuit mounted on board
		Press button switch
		Box
1	1 cm³	Syringe containing 0.72 cm³ epoxy resin

SET VALUES

Instrument	Set value	Calibration		Range of Tolerance*
		marked divs.	Sub divs.	
Measuring cylinder	53 cm³	10 cm³	x 1	52 — 54
Forcemeter	17 N	10 N	x 2	16 — 18
Lever arm balance	119 g	50 g	x 2	118 — 120
Ruler	68 mm	1 cm	1 mm	67 — 69
Thermometer	48°C	10°	x 1	47 — 49
Stop-clock	437 s	5 s	x 1	436 — 438
Ammeter	0.27 A	0.1 A	x 0.02	0.26 — 0.28
Voltmeter	1.3 V	1 V	x 0.01	1.2 — 1.4
Syringe	0.72 cm³	1 cm³	x 0.01	0.71 — 0.73
Manometer	220 cm H_2O	1 cm	x 1	21.0 — 23.0

*Pupils awarded maximum marks if reading set values within range of tolerance of instrument

Figure 3   Using measuring instruments

Source: DES/APU, 1984 (Reproduced with permission)

Sub-category   Estimating physical quantities.

Sub-category   Following instructions for practical work.

Category 3 (figure 4) observation. The act of observing is closely integrated with interpretation and is dependent upon existing scientific knowledge. Observation is not perceived as a theory-free activity.

Sub-category   Making and interpreting observations.
Example:

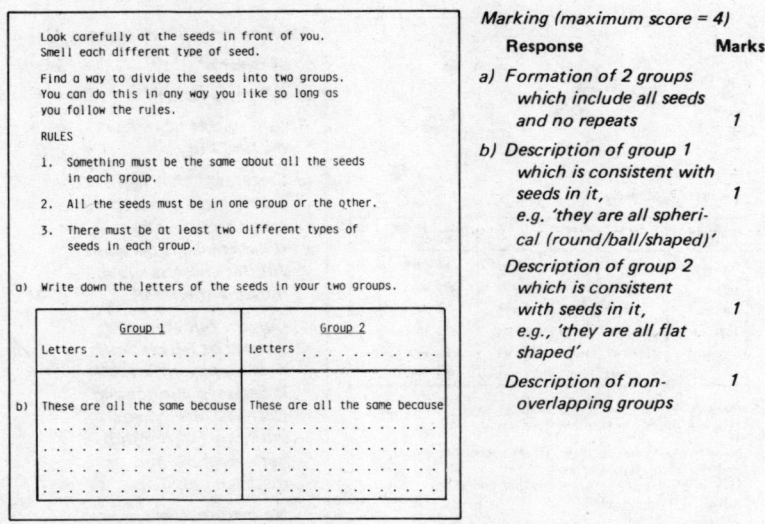

Figure 4   Making and interpreting observations

Source: DES/APU, 1984b. (Reproduced with permission.)

Category 4 (figure 5) interpretation and application.
Here, the skills of 'finding patterns in information, making predictions, judging the consistency between evidence and inference, drawing conclusions, giving explanations and suggesting hypotheses' (DES/APU, 1984b) are included.

Sub-category   Interpreting presented information. In this sub-category, there is an attempt to ensure that answering the questions depends on the use of the skill and not on the recall of knowledge of particular facts or concepts.
Example:

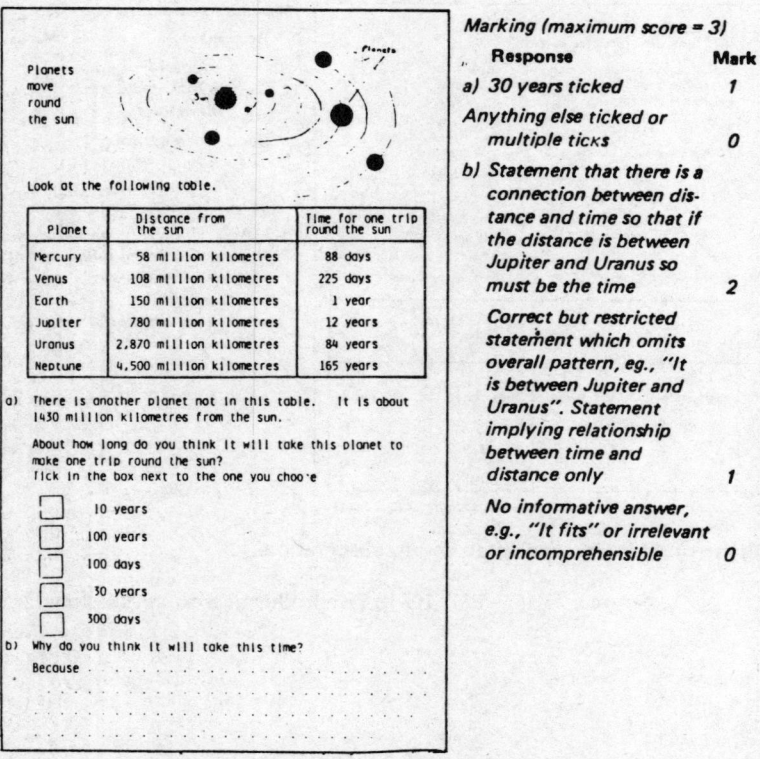

Figure 5   Interpreting presented information

Source: DES/APU, 1984b. (Reproduced with permission.)

Sub-category   Applying science concepts to make sense of new information.

Here, there is a requirement of pupils 'to recall and use relevant science concepts and knowledge in making predictions or suggesting explanations' (DES/APU, 1984b).

Category 5 (figure 6) planning investigations.

This category is concerned with children designing experiments to test out their ideas, taking into account variables and deciding on the observations and measurements that need to be made.

Sub-category   Planning parts of investigations.

Questions are asked on one part of the activity associated with planning an investigation.

Tim built a 'wall' using boxes as bricks, like this:

Flo built her wall from the same kind of boxes, like this:

They wanted to see which wall was the stronger. They decided to do this by rolling balls along the floor to hit the walls. To make this a _fair test_ they must keep some things the same for both walls.

Write down _3_ things that you think should be kept the same:

1. . . . .
. . . .
. . . .
2. . . . .
. . . .
. . . .
3. . . . .
. . . .
. . . .

Marking (maximum score = 3)

One mark each acceptable different response to a maximum of 3

Examples of acceptable responses:
— *Roll the balls — from the same distance away from the walls*
— *with the same amount of force when rolling/ strength of pushing/same speed/distance*
— *test in same place/room for both walls*
— *hit the wall in the same position*
— *same ball/weight or ball/ size of ball*
— *same number of balls used*
— *Walls — same height*
— *same number of boxes*
— *same height of boxes*
— *boxes same distance apart*
— *same thickness of wall*

Examples of unacceptable answers:
— *build walls in same way*
— *bricks in same pattern*
— *instructions (e.g. do it three times)*
— *alternative tests (e.g. sit on them)*

Figure 6   Planning parts of investigations

Source: DES/APU, 1984b. (Reproduced with permission.)

Imagine you are stranded on a mountainside in cold, dry, windy weather. You can choose a jacket made from one of the fabrics in front of you.

This is what you have to find out:

> Which fabric would keep you warmer?

You can use any of the things in front of you. Choose whatever you need to answer the question.

You can:

● use a tin instead of a person

● put warm water inside to make it more lifelike

● make it a 'jacket' from the material

● use a hair dryer to make an imitation wind (without the heater switched on, of course!)

Make a clear record of your results so that someone else can understand what you have found out.

Survival

5 cans

2 same dimensions, aluminium A, B

1 same dimension, plastic E

1 same height but larger diameter, aluminium C

1 same diameter but shorter, aluminium D

2 cover for container with bung and

110°C thermometer

Rubber bands

Pins

Sellotape

Scissors

Electric kettle

2 Measuring cylinders 100 cm³

Sheets of blanket

Sheets of plastic

Hairdryer

Ruler (30 cm)

Clamp for hairdryer

Graph paper

Cold water container

Paper towel

N.B. from admin. box

{ Pencil
Pen
Rubber
Stop-clock }

Notes for administrator

Put out sufficient fabric for each test situation i.e., ≃ 3 sheets each.

Remember to put out stop-clock, etc., from the admin. box.

Heat the water in the kettle prior to testing.

In case of breakage of thermometer please borrow one from your own school or the one you are visiting.

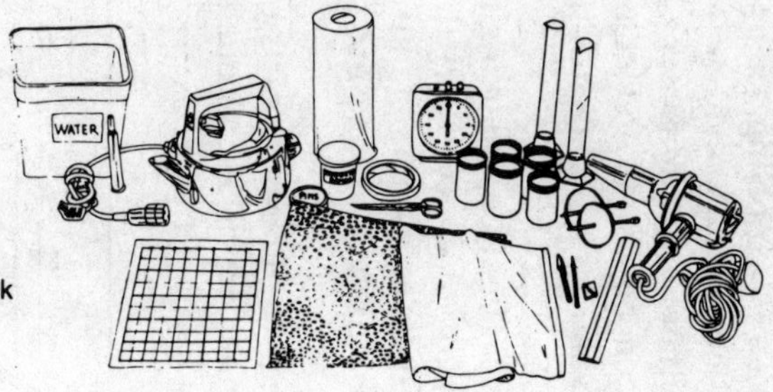

Figure 7 Performing investigations.

Source: DES/APU, 1985b. (Reproduced with permission.)

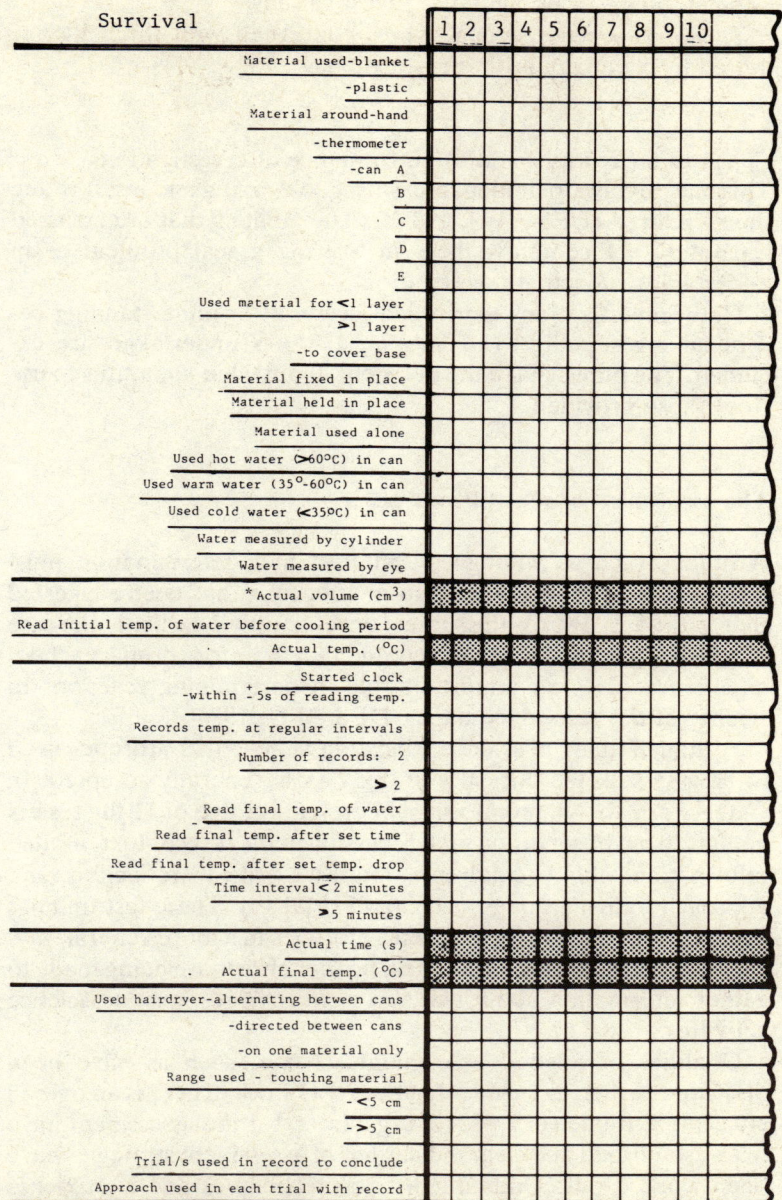

Figure 8 Checklist for investigation

Source: DES/APU, 1985a. (Reproduced with permission.)

Sub-category   Planning entire investigations.
These are open-ended activities concerned with the skills required in planning entire investigations.

Category 6 (figures 7 and 8) performance of investigations.
This may involve 'planning, measuring, observing and interpreting their findings' (DES/APU, 1984b). The abilities that are assessed separately are combined here in 'solving a real problem or in undertaking an actual investigation'.

Here, investigations that relied on 'recall or understanding before an investigation' (DES/APU, 1987) is undertaken are excluded. The pupils make the decisions about what apparatus to use from those provided.

The assessment of scientific attitudes

Attitudes such as willingness, interest and determination must influence how children perform in science. It has been suggested that 'attitudes towards the subject develop rather earlier in the case of science than for other subjects and that therefore primary school experience plays an important role in determining reactions to science at the secondary stage' (DES/APU, 1984b).

Although many in science education believe that attitudes need to be assessed, the difficulty of this task is generally accepted. In Category 6, above, monitoring children at the age of 11 the testers gather, 'from observation and discussion in the test context, limited information about children's interest, critical reflection and approach to living things' (DES/APU, 1984b). These test findings do not, however, reflect children's attitudes in a more general way. By employing various techniques, further efforts are being made to gather information about children's effective response to science activities.

Qualities of a good assessment scheme, such as have been discussed earlier, are evident in the APU work. There is an overall conceptual framework where 'organisers' for the assessment have been established and clear indications of assessment strategies have been worked out. Undoubtedly, the techniques and approaches developed by the APU to assess populations of pupils and gain useful information about the skills, concepts and attitudes of students could be adapted to produce vital information on any record of achievement or profile produced for a student.

The APU uses two kinds of practical test, both of them administered by trained testers and involving detailed mark schemes or checklists. To assess 'making and interpreting observations' groups of eight pupils are given materials or events to observe before completing written response sheets. For the assessment of 'Using apparatus and measuring instruments' groups of eight pupils are involved in a variety of tasks including estimation of physical quantities, reading measuring instruments, performing standard techniques and following instructions. These activities take the form of a practical circus, around which the pupils move. Further details of the organisation and administration of the 'observation' and 'measuring' circuses are given in *Practical Testing at Ages 11, 13 and 15* (DES/APU, 1985a). At age 11 the assessment of 'Using apparatus and measuring instruments' has been assessed in conjunction with the 'Performance of investigations' (although this is unlikely to be the case in the next survey at age 11 to be conducted in 1990). The 'Performance of investigations' is assessed by individually presented practical tests, a form of presentation which some feel is unlikely to commend itself to teachers. The APU is, however, currently involved in research activities investigating various aspects of practical assessment, some of which might more directly address the problems associated with classroom assessment.

Techniques for the Assessment of Practical Skills in Foundation Science (TAPS)

TAPS research is based in Glasgow and is funded by the Scottish Education Department (SED). The two phases of research were conducted during 1980–83 and 1983–86.

This work has focused upon what Bryce and Robertson (1985) have called 'the non-trivial aspects of practical skills as they manifest themselves in the classroom'. The system is criterion-referenced. The schedule of skills to be evidenced has been devised in accordance with the ways in which teachers think and act in the school

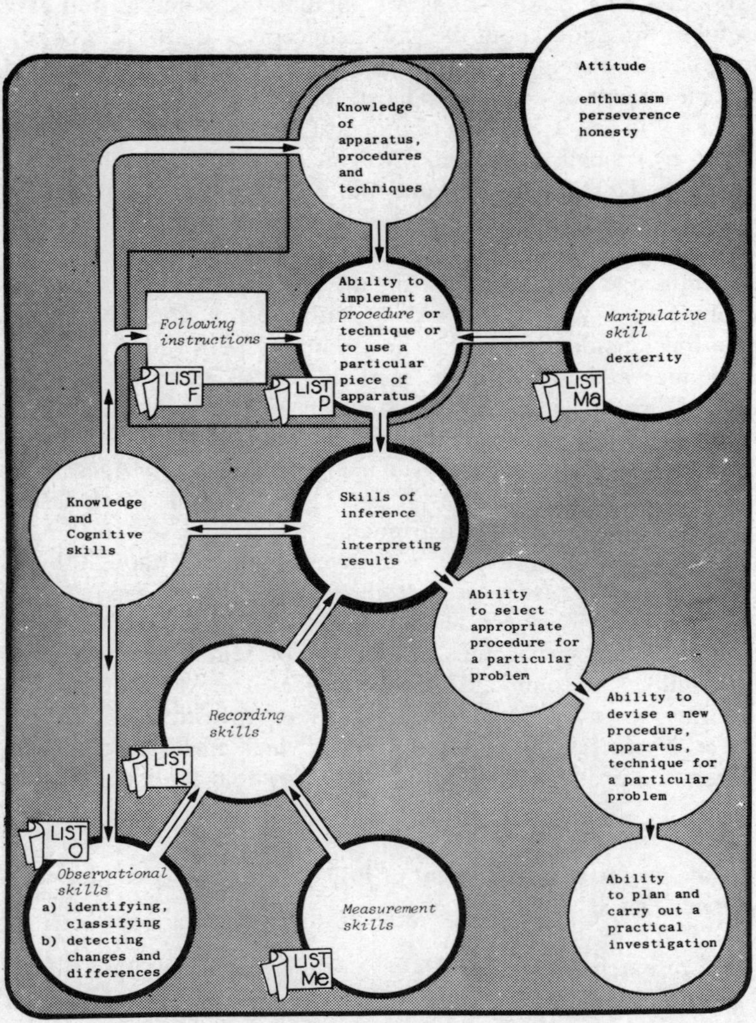

Figure 9   A conceptual analysis of the skills of practical science.

Source: TAPS, 1983. (Reproduced with permission.)

laboratory and how they talk to their pupils about practical work. The skill categories conceived were matched to coherent groupings of specific skill objectives, as identified from teaching materials. The six categories of Scottish Standard Grade (Foundation) Practical Skills are: Observational skills (O), Recording skills (R), Measurement skills (Ms), Manipulative skills (Ma), Procedural skills (P) and Following instructions (F). An additional two categories of skills are required at General level: Inference skills (In) and Selection of procedures (Sp) (selecting appropriate procedures for particular problems).

For the Standard Grade Foundation levels skill requirements are short, highly structured practical test items. At General levels, skill requirements of interpreting observations and results have led the research team to develop sequential sets of tasks that are also highly structured. The use of process checklists requiring the presence of an observer have been limited to a small number of basic laboratory procedures such as using a bunsen burner or thermometer. For achievement (or lack of it) there is a reliance on product-based paper and pencil or 'endcheck' test items. The 'stations' approach is regarded as a significant factor in influencing teachers to conduct practical assessments. This approach has been regarded as a worthwhile first strategy.

The TAPS team produced programmes of both the tasks and the skills that were to be used for the assessment. The coherent working scheme that emerged is illustrated in figure 9. The practical skills objectives were refined in the process of generating test items and in setting criteria for satisfactory performance; figure 10 shows the 46 objectives that are listed under the six skill categories. A summary of the detailed objectives is given in table 2.

The 315 items are offered as exemplars to assess the practical skills of the core areas of the Scottish Foundation Science course. Each item planner is devoted to an individual item or to a group of related items. Each item described in the Teacher's Guide has been designed to be used with its corresponding Pupil's Assessment Card. Each card contains one or more of the paperwork elements necessary for the presentation of the test item: instructions to the pupil, pupils' answer format with space for their name and, where required, instructions to the teacher and a detailed checklist. An additional element – an Endcheck Record Card – is supplied for endcheck items.

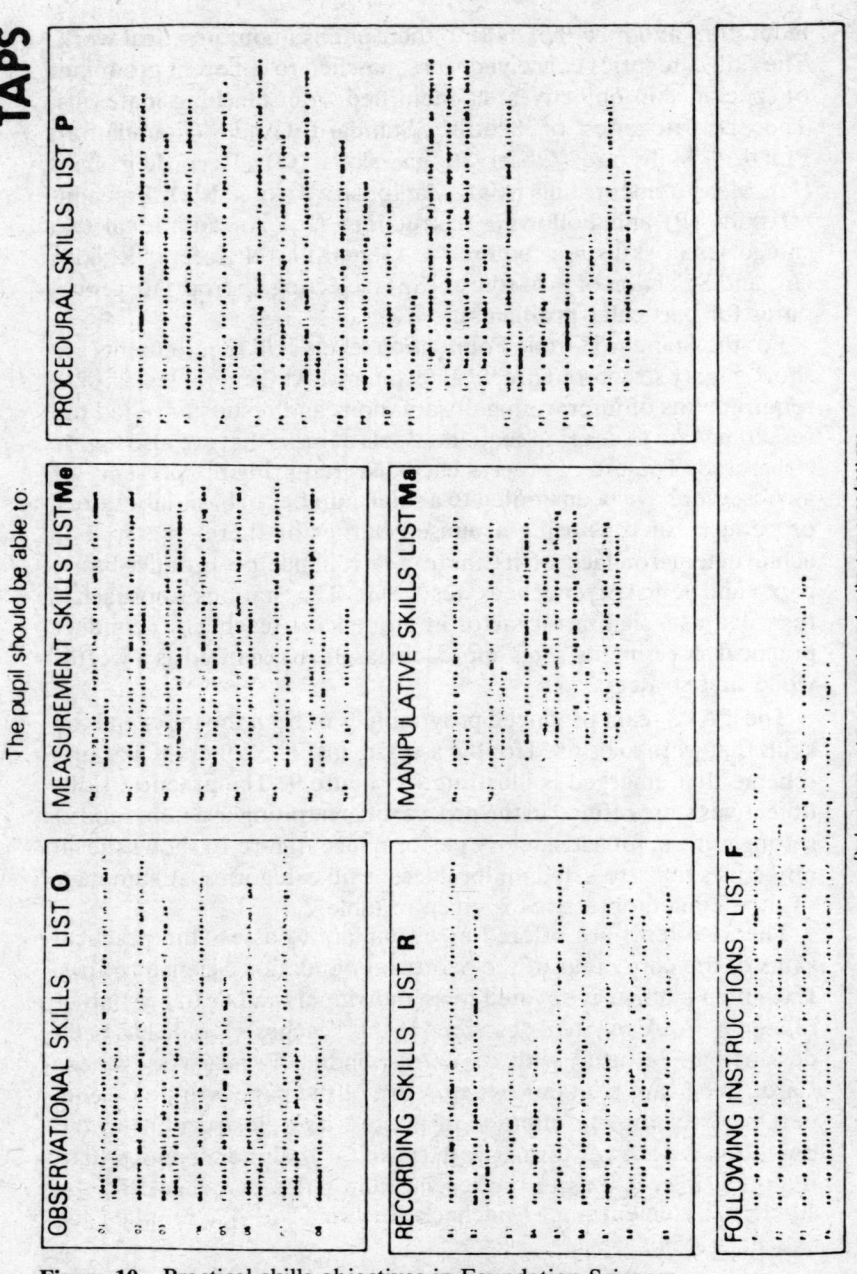

Figure 10 Practical skills objectives in Foundation Sciences.

Source: TAPS, 1983. (Reproduced with permission.)

Table 2 *A summary of the detailed objectives in the six skill areas*

Skill area	Ability to . . .
O Observational skills	identify, classify, detect changes and differences . . .
R Recording skills	present findings using spoken and written summaries, diagrams, tables and graphs . . .
Me Measurement skills	read scales, measure and supply units, estimate values . . .
Ma Manipulative skills	handle apparatus, coordinate movements, adjust and control, assemble and attach . . .
P Procedural skills	choose and use particular pieces of apparatus, implement a procedure or technique . . .
F Following instructions	follow demonstrations, spoken, written, or pictorial instructions . . .

Source: TAPS, 1983. (Reproduced with permission.)

All items are scored on a pass/fail basis. On each item planner, information is given for the scoring of pupil responses together with 'possible' and 'criterion scores'. In some items there is only one correct response, in others, every requirement must be met for students to achieve a pass. There are cases where simple slips may occur and pupils may fall short of a perfect score (e.g. as in item 02.5 'U-tube levels' where the possible score is a 5 and the criterion score is 4: pupils must achieve at least 4 to pass). In a process checklist item, prompts may be given if certain actions are not carried out spontaneously by the pupil.

In testing skills for the category 'Following instructions', a series

Table 3 *F objectives and test items*

		Objective			
Item number	Item name	F1	F2	F3	F4
.1	Assemble the apparatus			√	
.2	Assemble the circuit	√		√	√
.3	Assemble the molecule		√	√	√
.4	Battery hydrometer	√		√	√
.5	Combination locks		√		√
.6	Fix the tube	√		√	
.7	Folding paper	√			
.8	Lights				√
.9	Mixing solutions			√	√
.10	Mouldy bread		√		√
.11	Placing cards		√	√	√
.12	Plastic meccano		√	√	√
.13	Pooter			√	
.14	Pulley system	√		√	√
.15	Safe cracking		√	√	√
.16	Stapling	√			
.17	Syphoning	√			
.18	Using a map		√		√
		7	7	11	12

Source: TAPS, 1983. (Reproduced with permission.)

There is one sheet for each skill area together with one summary sheet.

of stages needs to be followed. Each of the F test items may be used to assess one or more of the four F objectives (see table 3).

Each of the items (.1 to .18) appears on a separate teacher's item planner (TIP) sheet. Details of F1.2, F3.2 and 4.2 ('Assemble the Circuit') are, for instance, given on one TIP sheet. The most difficult form of presentation is F4 where a list of written instructions is given.

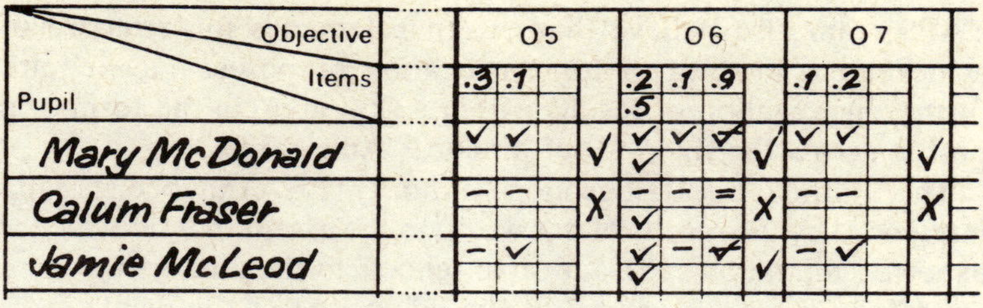

	Objective		O 5			O 6			O 7		
Pupil	Items		.3	.1		.2 .5	.1	.9	.1	.2	
Mary McDonald			✓	✓	✓	✓ ✓	✓	✓	✓	✓	✓
Calum Fraser			−	−	X	− −	−	=	X	− −	X
Jamie McLeod			−	✓		✓ ✓	−	✓	✓	− ✓	

Figure 11   An extract from the Observational Skills Practical Assessment record.

Source: TAPS, 1983. (Reproduced with permission.)

Practical assessment records

Practical assessment record sheets are included as part of the assessment packs for TAPS (see figure 11). If a student repeatedly fails to pass an objective and later succeeds in that same objective, that individual should be deemed to have achieved the objective.

In figure 11 the teacher has used two items tapping objective 05–05.3 (noting an item pass thus ✓) for Mary and item failures (−) for Calum and Jamie. This ✓ can be later transferred, with all the other summative assessments, to the summary sheet. Calum is judged not to have achieved the objective and an X is marked in the space to the right of the item pass/fail boxes.

Such detailed records of achievement can only help in the documentation of the strengths and weaknesses of students. Bryce et al. (1983) believe that the 'information on the record sheet lets the teacher see at a glance where particular pupils are in respect of their progress towards the acquisition of practical science skills'. The

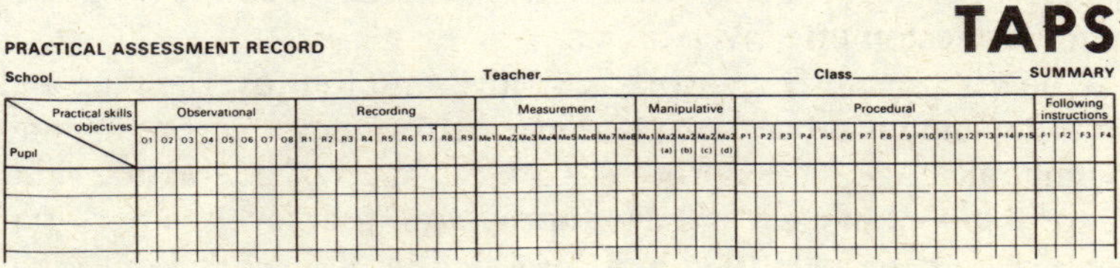

Figure 12   Practical assessment record.

Source: TAPS, 1983. (Reproduced with permission.)

TAPS team also believe that external demands for a record of achievement in itself or for certification purposes, may require further aggregation of results within a skill area 'in the form of a profile across the six kinds of practical skills'.

The Sheffield Local Education Authority (LEA) has been greatly influenced by the results of this work, to the extent that by 1988 it is expected that at least 21 out of 36 schools in the authority will be awarding a basic skills certificate based on TAPS. The certificate is seen as a motivating, achievable goal for all pupils.

The Oxford Certificate of Educational Achievement

The OCEA is being developed jointly by the University Delegacy of Local Examinations, the education authorities of Coventry, Leicestershire, Oxfordshire and Somerset and the Oxford University Department of Educational Studies. OCEA developmental work commenced in 1982. Piloting, which began in 1985, was completed in 1987 with a national launch in 1987/8.

The OCEA scheme consists of three different parts, the 'E-Component', the 'G-Component', and the 'P-Component'. The 'E-Component' consists of the record of results of external examinations. The 'G-Component' consists of graded assessments in four subject areas: English, mathematics, modern languages and science. Finally, the 'F-Component', which is a record of a student's achievements and experiences, is formative during the year at school and summative when leaving or changing school or college.

OCEA science provides a method of assessing any science subject because it does not prescribe content. Instead, it is based on the processes involved in investigative science. OCEA science does not assess knowledge, since this is tested in examinations taken at 16 plus, but recognises that without it, the other scientific processes cannot develop effectively.

The assessment framework is divided into four overlapping and interdependent processes: planning, performing, interpreting and communicating. These processes are made up of interacting skills (see table 4). For each of the four processes, four levels of performance have been identified and described by the general criteria.

The trials were intended to evaluate an assessment method known as coursework assessment. Evidence for assessment is drawn from observation of students, discussion of work with stu-

Table 4 *The assessment framework*

Process	Skills
Planning	Problem formulation Experimental design
Performing	Observing Manipulating Data gathering
Interpreting	Data handling Drawing conclusions Predicting
Communicating	Reporting Receiving information

(Adapted from the original and reproduced by permission of the University of Oxford.)

dents and marking the students' written work. The type of work assessed could include 'traditional' practical work, investigations, problem-solving and hypothesis-testing exercises, projects and written work. To exemplify the general criteria method employed for each of the four processes, 'Planning' will be discussed.

In defining criteria for the assessment of planning necessary to carry out our investigation or solve a problem, both the complexity of the task and difficulty of the concepts need to be considered. For this reason, criteria for levels 1 and 2 assume a simple task, with no requirement to control variables. Performance at levels 3 and 4 is demonstrable only on more complex tasks. Planning usually involves some discussion between students and with the teacher. If the students produce no plan of approach either on paper or in the carrying out of the experiment or investigation even after assistance, then they have not reached level 1. The four levels of performance by the student are:

Level 1 – produce a plan that would give some relevant results (awareness of the need for averaging and controls is not essential).

Level 2 – produce a plan which if carried out will answer a simple problem.

 – suggest or choose apparatus appropriate to the task.

Level 3 – design an experimental approach to a problem having some regard for variables and controls.
– modify the choice of apparatus if necessary.

Level 4 – understand the importance of recognising sources of error, and suggest consequent experimental modifications.
– change the approach if the first one is not successful.

Formulating problems, with appropriate methods of solving them, and suggesting further experimentation to investigate other aspects of the problem posed is indicative of performance at level 4 or above.

Criteria for the assessment of the other three processes have also been produced on similar formats.

The intention is to discover what children *can* do; therefore only positive statements are recorded. The research team accept that the criteria do not cover all parts of scientific activity, but hope that they include the most important aspects. Unlike with some schemes, a student must meet the requirements of each criterion without help, on several occasions and in different contexts before the success is recorded.

The OCEA team see the assessment framework as part of the approach to teaching and learning. The team is endeavouring to encourage motivation through student self-assessment and reflection on progress. For this purpose simplified criteria are being trialled in order to help students understand what is being assessed and to recognise their achievements. Students are encouraged to reflect on their progress and experiences by using diaries, checklists, questionnaires, or other appropriate materials. Discussion time, for students to air any problems, is also regarded as being important.

The 'P-Component' aims to offer opportunities for the development of educational processes which will:

- help students become actively involved in their own learning and development
- help students understand and accept responsibility for their own learning and development
- encourage the development and use of skills of self-assessment, the capacity for evaluating the importance of experiences and the ability and confidence to take responsibility for the planning and realisation of goals.

- help students develop a greater understanding of themselves and their relationships with other people
- encourage regular discussion between students and teachers, thus opening up possibilities for broader and deeper relationships.

Personally compiled records enable students to make statements about their interests, achievements and experiences. This development is supported by a parallel growth in social and personal education and tutorial courses to help students develop a realistic and positive concept of themselves, their achievements and their relationships with other people.

OCEA science is seen as part of the total OCEA package, which provides a personal record and a statement of public examination results as well as criterion-referenced assessments in four subject areas. The OCEA science certificate will include statements based on the criteria that have been met. It is believed that the substantial number of criteria applied throughout the 11–16 age range means that they will have a strong motivating potential for students of all abilities. In line with recent developments, it will also act as preparation for GCSE assessments and may be integrated into Mode 2 or 3 schemes.

The three major initiatives mentioned above have different reasons for the development of assessment techniques in practical science. The APU assesses students in order to establish national standards of achievement; nevertheless, the techniques developed lend themselves easily to being used by classroom teachers for their own purposes. Often pencil and paper techniques were used (as can be seen from the examples above), but experiments that reveal what the student can do were also developed. For OCEA, the assessment techniques arose from a desire to develop useful assessment methods in order to obtain information for profiling. In the OCEA scheme assessment and profiling are well integrated. The TAPS scheme, on the other hand, is primarily designed to assess practical skills and to provide a means of recording them.

The OCEA and the TAPS (for Scotland) may be regarded as important entities that schools are able to adopt in their entirety for the purpose of assessment and recording. The OCEA would be adopted by the whole school rather than by individual departments, since it involves assessment and recording in a broad spectrum of subjects. There again, all three (APU, OCEA and TAPS) are useful resources from which to select ideas for the development of

CLASS: _____

DATE: _____

TEACHER: _____

SKILL ASSESSED: _____

PRACTICAL USED: _____

Student's name	Check points									Level attained	Further comments

Figure 13   Checklist for GCSE assessment

better assessment and profiling schemes for practical science. No one scheme is 'better' than another, rather, particular aspects of one scheme may lend themselves more to the approach of a particular teacher. A probationary teacher, for example, may find the 'stations' approach a good first strategy before attempting to assess the whole class carrying out a more sophisticated experiment.

The three schemes described so far are prescriptive and do not represent the total picture of assessment developments in this country. The Dorset/Southern Regional Examinations Board (SREB) Assessment and Profiling Project is very different from what has been described above. This DES funded project is fundamentally different in that it is school-focused; each school designs schemes appropriate to its own purposes.

INTERNAL PRACTICAL ASSESSMENT APPROACH FOR GCSE

In the previous section we looked at three major assessment schemes; moving on from there we will now consider how the secondary school science teacher can put into practice some of the ideas discussed above when assessing practical abilities in science as part of GCSE courses.

Internal practical assessment now forms an important part of all GCSE science courses. An appropriate way to carry out such assessment is by the adoption of the checklist approach (see figure 13). Here, the skill area is sub-divided into appropriate points – the check points. Checklists are easy to devise and use and provide teacher security by adopting a tangible structure from which to work.

The final assessment can only be as good as the criteria and guidelines set by the examination boards for assessing. Skill areas have been hurriedly defined in many instances and criteria for the skill areas are often ill-conceived; nevertheless the teacher has to work within the confines set by the boards.

Examples of the skill areas identified by the biology syllabuses of two examination boards and how the assessment of two skills can be undertaken will be discussed next.

The Southern Examining Group (SEG)

This board has identified the following skill areas:

- following written and diagrammatic instructions
- handling apparatus and materials
- making and conveying accurate observations
- recording results in an orderly manner
- formulating a hypothesis
- designing an experiment to test a hypothesis
- carrying out safe working procedures

For each skill area, there are four levels of attainment: 0, 1, 2 and 3. For the skill area that requires the student to handle apparatus and materials, the description of the levels is as follows:

0 – Skill not demonstrated.
1 – Handled apparatus and materials with one major error, one major error and one or two minor errors, or several minor errors.
2 – Handled apparatus and materials with one or two minor errors.
3 – Handled apparatus and materials with no error.

    The experiment illustrated in figure 14 will be used to assess the skill. Firstly, the important check points are identified and a subjective decision on whether errors are major or minor is made (see figure 15). The relevant ticks (indicating satisfactory achievement of check points) or crosses (indicating unsatisfactory achievement of check points) are then translated to levels of attainment.

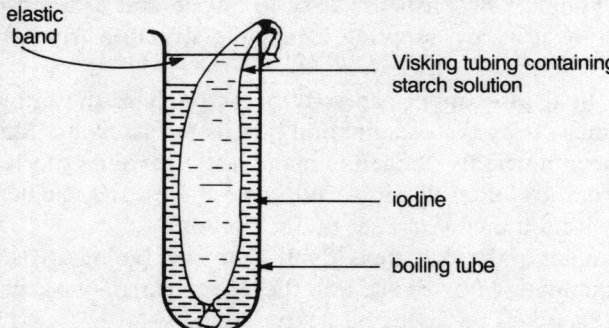

Figure 14   Visking tubing acting as a model of a gut

An Experiment to Demonstrate how Visking Tubing can be used as a Model of the Gut

Apparatus: Starch solution, elastic band, visking tubing moistened in water, 1 boiling tube, 1 test-tube rack, iodine solution, stop clock, spotting tile, 2 teat pipettes

Method:
1. To the wet visking tubing, tie a *tight* knot at one end.
2. Fill approximately ¾ of the visking tubing with a starch solution using a teat pipette.
3. Holding the other end firmly, rinse the outside of the visking tubing under a running tap.
4. Three-quarters fill the boiling tube with iodine solution.
5. Twist the top of the visking tubing and hold in place with an elastic band as shown in the diagram above.
6. Leave the boiling tube in the test-tube rack for 5 minutes. If the iodine turns blue/black, repeat this procedure.
7. Record what happens at the end of 5 minutes.
8. Add 1 drop of starch solution to a well on the spotting tile. To this, add 1 drop of iodine solution.
9. Record what happens.

The Midland Examining Group

This board has identified the following skill areas:

- following instructions
- handling apparatus and materials
- observing and measuring
- recording and communicating
- interpreting data
- experimental design/problem solving

The levels of attainment are High (H), Intermediate (I) and Low (L). Each level is further divided, thus:

$$H \begin{cases} 9 & \text{high} \\ 8 & \text{intermediate} \\ 7 & \text{low} \end{cases}$$

$$I \begin{cases} 6 & \text{low} \\ 5 & \text{intermediate} \\ 4 & \text{low} \end{cases}$$

Student's name	Check points					Level	Comments
	one drop of iodine added to starch / one drop of starch (MINOR ERROR)	Visking ¾ full of starch (MINOR ERROR)	Visking twisted and tied with elastic band to boiling tube (MINOR ERROR)	Rinsed visking (MAJOR ERROR)	If iodine went blue /black – repeated correctly (MAJOR ERROR)		
Allen, Anthony	✓	✓	✓	✓		3	
Christopher, Alex	✓	✓	✓	✓	✓	3	
Elmes, Harry	✓	✓	✗	✓	✓	2	
Hassan, Mehmet	✓	✓	✓	✓	✗	1	

Figure 15 Checklist for the ability 'to handle apparatus and materials'

$$L \begin{cases} 3 & \text{high} \\ 2 & \text{intermediate} \\ 1 & \text{low} \end{cases}$$

The board recommends appropriate experiments that may be used to attain information for each skill. For the skill area 'observing and measuring' the levels of attainment are defined as follows:

H – Handles quantitative work with confidence. Uses measuring instruments correctly and relates readings to quantities. Observations are detailed and accurate.

I – Quantitative observations are usually accurate in most features, but quantitative observations lack detail and discrimination.

L – Able to make detailed and relevant observations and measurements given considerable assistance.

The osmosis experiment in figure 16 is used to derive information for this skill area. The appropriate box on the checklist (figure 17) is ticked to identify the level attained. In this checklist, check points are grouped together to identify particular levels. For each level check points can quite easily be separated if this is found to be more preferable by the assessor.

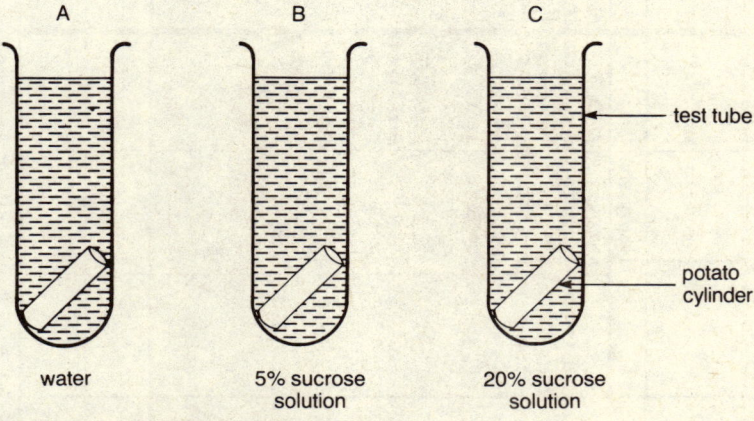

Figure 16   Osmosis in potato tissue

Experiment to Demonstrate Osmosis in Potato Tissue

Apparatus: Water, 5% sucrose, 20% sucrose, three cylinders of potato (approximately 4-cm lengths), 3 boiling tubes, test-tube rack

Level	L			I			H		
	Help given			No help given					
Check point	L	I	H	L	I	H	L	I	H
	Notes changes in fatness or size.	Notes changes in fatness and size and measurements recorded.	Notes changes in fatness and size correctly and mathematically with small error. Does not mention size increase or decrease.	Notes changes in fatness or size.	Notes changes in fatness and size. No measurement recorded.	Notes changes in fatness and size correctly and mathematically with small error. Does not mention size increase or decrease.	Notes changes in fatness. Small arithmetic error or no mention of size increase or decrease.	Notes changes in fatness. Small arithmetic error. Some indication of size increase or decrease.	Notes changes in fatness. Gives arithmetic difference precisely. Indicates size increase or decrease.
Student's name									

Figure 17 Checklist for 'observing and measuring'

Method: 1. Place one 4-cm potato cylinder in each of three boiling tubes.
2. Pour water in A, 5% sucrose in B and 20% sucrose in C, as shown in the diagram above.
3. Stand in a test-tube rack for one or more days (depending on when it is most convenient to go back to the experiment).
4. Remove each cylinder by pouring the contents of each boiling tube into a clean beaker.
   Measure each cylinder and note any change in length.
5. Record any differences in firmness.

As can be seen, different checklists can be produced to suit the skill being assessed, the board's requirements and individual preferences. It is not the intention here to give long lists of checklists and practicals for teachers carrying out GCSE assessments; the approach is to indicate the methodology involved in producing the checklists.

By using checklists, a high degree of comparability can be achieved between schools and individual teachers within schools. This presumes that the same checklists are used which at present is highly unlikely; none the less, comparability is only valid if skill areas are meaningfully defined to assess the educational process or product and are fair to students. Examination boards must take care that all is not sacrificed for the sake of standardisation. If the skill areas are ill-conceived then the value of the assessment is diminished. An example of this failing is where certain examination boards decide on the level of attainment merely by the number of student errors, often not taking into account students correcting those errors. This method has to be educationally unsound.

There are many other questions to which teachers must apply themselves before they carry out any internal assessment. Some of the more important ones are highlighted below.

Questions to be considered

How many skills are assessed per lesson?

There may be recommendations from the board on the number of skill areas that should be assessed on any one occasion. It must be quality though, rather than quantity, that is of prime importance. From the same practical session it may, nevertheless, be possible to assess one skill (for example, 'handling apparatus') through observation and another (for example, 'interpreting data') from a written account. It is not advisable to assess too many skills requiring observation at any one time, especially if large groups are to be assessed. If team teaching is involved it may be possible for each teacher to assess different skills.

How many students are to be assessed during one practical session?

Initially it may be a good idea if a teacher assesses small groups or only part of a class in order to gain confidence and experience in tackling larger numbers or whole classes. If only some students are assessed then the teacher has to make decisions about how to treat the rest of the class. Teachers can overcome the problem by not informing students in advance of whom is to be assessed and every student undertakes the same experiment. This avoids having to give some students different work or having to recruit a colleague to 'help out' with those not being assessed.

What conditions need to be created in the laboratory?

It is interesting that examination boards state that assessments must be carried out as part of the teaching process and formalised conditions must not be created. Yet boards require teachers to ensure the individual contribution of each candidate. Typically, they do not state how this is to be achieved, and although there is no requirement by GCSE boards to assess this attitudinal skill in science, talking and working in groups or pairs can encourage co-operation.

Faced with the problem of ensuring the individual contribution of students' work, teachers must decide on their approach. If the teacher allows normal classroom interaction between students then only a very small number can be assessed on any one occasion. It is difficult to observe large numbers of students and know the exact contribution of any individual. When it comes to the skill areas 'handling apparatus', 'following instructions' and 'working safely' larger groups can be tackled. On the other hand, it is difficult to ensure individual pupil contribution in 'interpreting data', 'experimental design/problem-solving', 'formulating a hypothesis', etc. if pupil discussion is allowed.

The stations approach can also be employed where students are carrying out separate experiments in different parts of the laboratory. This is very demanding of technicians' time, but can reduce the chances of cheating. Also, if help is given by the teacher to one group of students, there is not the concern that other students will observe and copy the teacher.

For some experiments, the teacher may be helped if students are asked to indicate by raising their hands when particular points in the experiment have been reached. For example, if the experiment involves food tests, it is possible that different colours may be obtained from the tests. This method enables the teacher to check the colour of the reaction against the students' observation.

Class experiments are not always needed. If an experiment involves expensive resources or if there is a shortage of equipment, skills such as 'observation' can also be assessed from a teacher demonstration or even a film of the procedure.

How is standardisation ensured?

Standardisation of work between schools is achieved through moderation (see Chapter 2) and this is the responsibility of examination boards. Internal standardisation is a departmental matter that may be achieved through departmental discussion on approach and use of practicals. A good way of reaching a common approach is through teachers assessing groups together. Such team teaching will ensure that individuals are attuned to each others' thinking even though it need only be done on a small number of occasions. It is, of course, important that the methodology used by teachers is the same. Checklists, such as those described previously, will ensure a high level of comparability.

The schemes described in the first part of this chapter should provide the teacher with some tangible ideas on implementing internal assessment, despite the fact that in looking at GCSE assessment, it has to be acknowledged that there is always the added frustration for teachers of having to work within the confines of the examination boards' guidelines, even if those guidelines are impracticable or even educationally questionable.

Chapter 4

Profiling

This chapter commences with a general discussion on profiling, leading on to a consideration of the different possible types of recording. Some mention is also made of how practical skills in science can be recorded linking up with what was said in Chapter 3. The chapter ends with a description of how a notional system of profiling can be operated.

BACKGROUND

Educational reports during this century have been paving the way for the introduction of profiles. The idea that the recording of educational achievement ought to span a much wider spectrum than that encompassed within more traditional assessment procedures, however, has grown rapidly over the last decade from a couple of isolated initiatives into a national policy commitment.

The story of profiling proper begins in the early 1970s with two different initiatives, one in the south west of England, one in Scotland. In Swindon, Stansbury and others had launched the RPA, which was taken up enthusiastically by many schools. Its aim was to provide every participant, no matter what their ability, with a personally compiled record of events, achievements and experiences. The Scottish initiative attempted to fit into existing assessment and certification practices by adopting a norm-referenced approach. The RPA helped to 'provide the 60 per cent or so of Scottish school-leavers then disenfranchised by the existing certification system with a detailed record of their achievement in every

70

area of school life' (Broadfoot, 1984). Since these early attempts, profiling schemes have mushroomed in number and also diversity.

Three stages have been identified in the development of profiles and records of achievement (Broadfoot, 1986). The first stage, from 1970 to 1980 – the 'mission' stage – was when the main principles were being set down. The second stage from 1980 to 1984 – the 'disseminated development' – was distinguished by a large number of small-scale initiatives, cross-fertilising to produce five or six generic approaches to recording achievement. The third stage, which exists now, is characterised by the development work becoming increasingly large-scale.

Broadfoot has observed that, increasingly, the efforts of individual institutions are being progressively overtaken by the local authorities and examination boards. In its publication *Records of Achievement: A Statement of Policy* (DES, 1984) the Government committed itself to a policy of providing all school-leavers with such a record by the end of the decade. It was for this reason that the DES provided just over £2 million over a three-year period to fund nine pilot schemes in England and Wales.

There are many profiling initiatives, and it will be interesting to see whether the DES will succeed in bringing all such record of achievement schemes within its national guidelines when it has drawn them up in late 1988. Broadfoot (1986) believes that 'if national guidelines are successfully imposed, the effect may be to dampen the grassroots enthusiasm'. Against this argument though, wide currency can only be achieved by the support of large consortia or examination boards.

There is now the danger that a hasty move may be made to institutionalise an ill-conceived or insufficiently developed scheme. To some extent the forcing through of GCSE and the assessment that accompanies it by Sir Keith Joseph in 1986 is an example of an over hasty attempt to impose a system regardless of teacher anxiety about the timing of its implementation. Broadfoot (1986) explains that once a system is adopted, it is difficult to change – like the examinations that were hastily developed in the nineteenth century and which are now proving difficult to remove.

Over the last century, external examination certification has dominated the assessment scene. Throughout this period of time there have been many demands for wider patterns of accreditation. Examinations are regarded as inaccurate and very limited in assessing what a pupil can do. Furthermore, they encourage extrinsic motivation rather than the desire to learn for its own sake, and they discourage co-operation between students and emphasise individual competition.

Broadfoot (1984) identified the following general aims of schooling:

- knowledge and understanding of self
- understanding of society
- development of interpersonal skills
- willingness to learn
- growing maturity
- preparation for effective community membership
- emotional and moral development
- commitment to democracy
- life enrichment.

In practice, it appears that pupils emerge from schooling lacking many of the personal and social skills they need. School is often perceived as irrelevant to real life and to the concern of young people. It also tends to create dependent, apathetic learners whose capacity for initiative, self-reliance, self-discipline and decision-making is stunted. All too often, it creates passive, alienated young people, lacking in the confidence to challenge what they see as their own powerlessness in a hierarchic and authoritarian society.

In 1971, the school-leaving age was raised to 16. The aims were to establish equally prestigious, but different, educational goals for the public examination courses and the new 'Newson-style' and later 'Rosla' courses (these were 'stop-gap' courses for students who would not normally be staying on) were established. Nevertheless, the superior standing of O-levels overshadowed the success of other courses at this level.

Greater youth unemployment and the economic decline in the seventies was instrumental in bringing about greater competition for qualifications and employment. Profiling, to a large extent, appeared to many in education as a means of providing an alternative form of certification, one which would encourage motivation among the dissatisfied, non-certificated groups and also promote the acquisition of more work-related skills and qualities and provide a better basis for selection.

The British Institute of Personnel Management (IPM, 1984) set out what it considers the definitive list of what employers look for in recruiting staff. It lists the following:

literacy
numeracy
communicative ability
organisation of work
ability to work with colleagues
ability to work with people in authority
analytical ability and problem-solving
judgemental and decision-making ability
adaptability
responsibility, self-awareness and maturity

If this is a valid list, then we disadvantage our pupils educationally and occupationally with our limited academic examinations and, since the qualities employers seek match very closely the outcomes we intend for our children's education, we must make revisions. Our assessment procedures must change and broaden in order to stress other educational outcomes. Assessments may reflect achievements that are part of a course, or those from extra-curricular and even extra-school activities.

Information obtained from assessments needs to be reliable, easily gathered, relevant and readily interpreted. The specific achievements of students need to be assessed and for this norm-referenced comparisons of students are inappropriate.

Profile records should be compiled over a long period of time, that is to say they should be formative in nature, incorporating the dynamic aspects of the teaching–learning process. From the diagnosis of individual needs and progress in the classroom they should advance through continuous guidance, pastoral and reporting functions and should culminate in, rather than emanate from, the need for a terminal certification (this would be the summative profile).

Nearly all profiles reflect the desire to present a comprehensive picture of the student. The ILEA report on the curriculum and organisation of secondary schools (1984) distinguishes four aspects of achievement:

(a) written expression, organisation of material, commitment to memory and similar academic achievements traditionally measured in formal examinations

(b) practical skills, the application of knowledge, oral and investigative skills (the application of the knowledge acquired under (a), only limited parts of which have traditionally appeared in formal assessments)

(c)   personal and social skills, communication and relationships, working in groups, responsibility and other such personal qualities not normally explicitly measured in traditional assessments

(d)   motivation and commitment, perseverance, self-confidence and self-image.

Similar broad areas that might be included in the profile have been set out for the Dorset/SREB scheme.

Area (a) above has traditionally been assessed in our schools, but if we regard areas (b), (c) and (d) as important educational objectives then achievements in these domains must also be recorded as part of the certification process if they are to receive the necessary emphasis in the curriculum.

The departure from written examinations and objective tests has, in some instances, led to novel ways of assessing. In England and Wales, and to a greater extent in Scotland, assessment techniques involving oral examinations and systematic observation, as well as graded tests and credit accumulation, student self-assessment and self-recording along with teacher–pupil negotiated assessment have all played a more significant part in school assessment. These methods of assessment are mentioned, to a greater or lesser extent, elsewhere in the text. Broadfoot (1986) thinks it significant that pupil-assessment, for example, 'which would have been unthinkable to most people (and certainly to politicians), even in the late 1970s, is in 1984 a central pivot of the policy on records of achievement'.

THE DIVERSITY IN PROFILING PRACTICE

At present, a series of government development schemes are being locally and nationally evaluated with a view to providing a national record of achievement scheme at the end of the decade. It is believed that the new records of achievement will address both the assessment-as-curriculum (the part played by assessment in the schooling process) and assessment-as-communication (the production of a summative report) functions of the profile.

A major division in the profiling movement is the priority given to the curriculum, as against the communication function of the recording system. At one end of the continuum of the recording system are those forms of recording that are subjective and person-

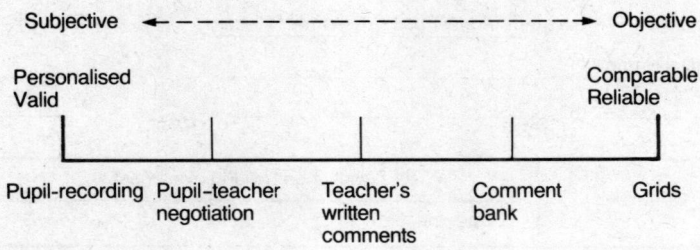

Figure 18 Types of record

Source: Broadfoot, 1986. (Reproduced with permission.)

al, with the main purpose of facilitating pupils' understanding of themselves and their own progress. At the other end of the continuum, the emphasis is on records that are fairly comparable and reliable; objectivity plays an important part in such records.

Pupil recording

Personal pupil recording (see figure 19) has been a significant development in the search for new forms of accreditation. The assessment is completely in the hands of the pupil. The emphasis is upon the students' image of themselves. This formative recording is entirely student centred with 'feelings, interests and personal development being of paramount importance' (Mortimore, 1984). The emphasis is on the development of a positive self image, of self confidence and of the skills of self-assessment as preparation for adult life. This form of self-assessment also has the benefits of not involving peer group comparisons and encourages discussion and good student–teacher relationships.

These records have the disadvantage of being perceived as being designed for non-examination students. Such records have little currency value in comparison with external examinations and assessment. It can also be argued that having no summative component is a weakness of such a system.

The use of grid recording

Early 'grid methods', such as the prototypes of the records of

Personal Pupil Assessment Record

Student's Name: _____   Year: _____

Tutor: _____   Class: _____

Qualities	Written evidence provided by student		
	Autumn	Spring	Summer
Determination			
Ability to communicate – written			
– oral			
Punctuality			
Attendance			
Reliability			
Concern for others			
Co-operation			
Ability to work alone			
Initiative			
Self-confidence			
Sociability			
Presentation of work			
Ability to discuss			
Adaptability			
General behaviour			
Maturity			

Other comments the student wishes to make:

_____

_____

_____

Students must refer to separate sheet describing above qualities.

Figure 19 Personal pupil assessment sheet

achievement of the Scottish Council for Research in Education (SCRE) profile, emphasised the comparability aspect. They provided 'grids' of descriptors of a wide range of qualities and skills against which assessments were made, usually on a four-point scale. Some of these grids were norm-referenced, and some more recent varieties emphasise criterion-referencing.

Profile grids (figure 20) present a series of hierarchical statements relating to a student's mastery (usually over five or six levels) of a specific skill. The grid usually consists of boxes to be ticked set against the appropriate level of achievement, or of bars that can be shaded to the relevant level. Mortimore (1984) claims that in the Avon survey 'most profiles contained grid elements, frequently based on the FEU/CGLI approach'. It also appears that profile grids were less common in schools that had developed a system to suit their own organisation than in schools new to profiling who accepted ready-made systems. Some of the schools used formative grids in conjunction with summative criterion checklists.

The progress grid sheet in figure 20 provides four categories of basic abilities covering a range of skills that can be developed throughout the entire course. For each of these categories five levels of attainment are defined. As each ability is demonstrated the bar is dated and shaded in to the level achieved.

Progressive grids can be motivating to students and are able to indicate clearly what a student has achieved across a wide range of skills and personal qualities and also what needs to be done to reach the next level. The presentation of information is compact and uniform and can be assimilated quickly (especially when shaded bars are used).

There are in-built dangers in such grids in that they encourage labelling and comparison between students, and employers and other users will only look at the top level attainable. From a practical level they are flawed because the steps in the grid are often uneven across a parameter, or the distance from one level to the next is so great as to be unattainable in practice.

Comment bank recording

More recently, in an attempt to overcome the mechanistic impression of grids, support has been accumulating for the idea of a 'comment bank'. A bank of statements is compiled (usually

SPECIFIC APPLICATIONS OF ABILITIES

COMMUNICATION	TALKING AND LISTENING		Can make suitable responses when spoken to
	READING AND WRITING		Can read short sentences
	USING SIGNS AND DIAGRAMS		Can recognise everyday signs and symbols
SOCIAL ABILITIES	WORKING IN A GROUP		Can cooperate with others when led
	WORKING WITH THOSE IN AUTHORITY		Can follow instructions for simple tasks and carry them out under guidance
	WORKING WITH CLIENTS		Can help someone to carry out a client's request
PRACTICAL & NUMERICAL ABILITIES	USING EQUIPMENT		Can use equipment safely to perform simple tasks under guidance
	CONTROL OF MOVEMENT		Can lift, carry and set down objects
	MEASURING		Can sort objects into sizes
	CALCULATING		Can count
DECISION-MAKING ABILITIES	PLANNING		Can describe the sequence of steps in everyday tasks with prompting
	SEEKING INFORMATION		Can ask questions
	COPING		Can cope with everyday activities with help
	ASSESSING RESULTS		Can seek advice about own performance

N/O — No opportunity to assess.

Name of Trainee/Student

Name of Scheme

Period covered by this review

Figure 20 Grid and profile

PROFILE REPORT

No.—

ATTAINMENTS IN BASIC ABILITIES

			(High Level)	★
Can hold conversations with workmates. Can use telephone to take messages	Can follow and give simple descriptions and explanations	Can communicate effectively with a range of people in a variety of situations	Can present a logical and effective argument. Can analyse others' arguments	
Can read and write messages	Can follow and give straight-forward written instructions and explanations	Can use instruction manuals and can write reports describing work done	Can select and criticise written data and use it to produce own written work	
Can explain signs and symbols	Can after guidance make use of basic graphs charts tables drawings etc	Can interpret and use basic graphs charts tables and drawings	Can construct graphs etc and extract information and use it to produce own written work	
Can work with other members of group to achieve common aims	Can understand own position and results of own actions within a group	Can be an active and decisive member of a group	Can lead a group	
Can follow instructions for a simple task and carry it out independently	Can follow a series of instructions and carry them out independently	Can carry out a series of tasks effectively given minimal guidance	Can assume responsibility for delegated tasks and take initiatives	
Can carry out client's requests under supervision	Can carry out client's requests without supervision	Can recognise client's needs	Can anticipate and respond to client's needs	
Can use equipment safely to perform several-step tasks after demonstration	Can select and use suitable equipment and materials for the job without help	Can set up and maintain equipment	Can identify and remedy common faults in equipment	
Can use everyday tools	Can use tools to produce work to given limits	Can use tools to do fine and/or complex work	Can perform tasks requiring a high degree of control	
Can read simple scales and dials	Can measure out specified quantities of material by length weight etc	Can read precision instruments	Can set up and use precision instruments to make accurate measurements	
Can add and subtract whole numbers	Can use × and ÷ to solve whole number problems	Can add and subtract decimals and simple fractions	Can multiply and divide decimals and simple fractions	
After demonstration can identify the sequence of steps in a routine task	Can choose from given alternatives the best way of tackling a task	Can modify/extend given plans/outlines to meet changed circumstances	Can create new plans routines from scratch	
Can find needed information when guided	Can use standard sources of information	Can assemble information from several sources	Can show initiative in seeking and gathering information from a wide variety of sources	
Can cope with everyday problems. Seeks help if needed	Can cope with changes in familiar routines	Can cope with unexpected or unusual situations	Can offer sensitive and effective help to the group	
Can assess own results with guidance	Can assess own results for familiar tasks without help	Can assess own performance and identify possible improvement	Can assess the group's performance and help to improve it	

N/O — No opportunity to assess.

Enter ★ in end column if high level is exceeded

Signed (Trainee/Student)

Signed (Supervisor/Tutor)

Date

(MK II)

Source: CGLI, 1982. (Reproduced with permission.)
This is a reduced copy of the original; normal size A3 – 297mm
(11.7m) × 420mm (16.5m).

between 10 and 30) relating to each area of student achievement. Teachers select pre-coded comments (A3, B4, C6, etc.) that are then processed and printed out by computer to read as a series of prose descriptions to each student.

Hargreaves (1986) uses examples from the draft personal record sheet currently being considered by one LEA to illustrate how the comment bank works. The tutor (in some cases with the student) rings or ticks an appropriate category under headings like 'attitude to people', 'attitude to activities', etc. Under two sub-sections of 'attitude to people', for instance, the tutor must ring at least one of the following:

P3A   He/she has a cheerful personality
P3B   Considerable concern for others has been shown
P3C   His/her confidence and assurance enable him/her to relate to other people well
P3D   The openness of his/her relationships enable him/her to solve conflict situations sensibly
P3E   He/she has shown himself/herself to be a responsible person
P3F   He/she is well mannered and courteous
P3G   He/she has a lively sense of humour
PD1   corresponds to a pupil who finds it very difficult to maintain relationships with (a) adults (b) pupils
PD2   corresponds to a pupil who refuses/is unable to relate to others despite encouragement
PD3   corresponds to a pupil who has formed no lasting relationships
PD4   corresponds to a pupil who is aggressive or lacking in self control
PD5   corresponds to a pupil who has shown neither an inclination, nor a capacity, for leadership.

The final continuous summative statement is an assemblage of encircled, numbered categories. Hargreaves claims that 'it is a diagnosed grid, no less'.

Comment banks have the advantage of avoiding norm-referenced grading and labelling of students (such as those involving a 1–5 scale or grade A–E), but ensuring a greater degree of comparability than is possible with free response. Mortimore (1984) claims that 'a potentially wide and accurate range of descriptors is available which is likely to provide positive comments meaningful to both student and potential user'. Without the use of coding for computers and word processors, this is a time-consuming approach for both the teacher involved and whoever provides clerical assistance.

With such records, the user can only discover what the students have not achieved by having access to the complete list of comments. Teachers using the comments would have to make themselves very familiar with the list and even then there is a danger that 'middle of the road' comments will be more widely used than others. Competencies such as numeracy are not easy to portray by means of comment bank. Hargreaves also expresses concern that such methods should be used to provide objectivity and comparability in the whole range of personal and social achievement.

Despite its somewhat limited use, this is a recording method that appeals to many teachers because of its ease of application.

Negotiated recording

Worries about comment banks have been instrumental in the development of the concept of negotiated records. In this system the teachers and pupils regularly discuss the pupil's progress, and in the course of these discussions set targets for the next set of learning objectives. Targets may be academic, social or personal. Burgess and Adams (1980) envisage that such a formative procedure would extend throughout a particular course and terminate in a more summative document, which would contain only that information that the pupils themselves wish to have included (see figure 21). The advantages and disadvantages of personal pupil recording have already been discussed.

Written comments ensure a more sensitive approach, depending of course upon the quality and sympathetic nature of the comment. However, this method could be most appropriate as indicated above, to stimulate effective discussion about the student and lead to worthwhile guidelines.

The use by a school or department of one method of recording should not exclude other types being used concurrently. It may be, for example, that a science department wishes to use progress grids in recording manipulative skills and also employ pupil–teacher negotiation, with a free comment format, for attitudes.

The methods we have looked at are not intended as a comprehensive study of available schemes of recording (these are far too

Negotiated Record Sheet

Student's Name: _____   Year: _____

Tutor: _____   Class: _____

Qualities	Student's statement on how the qualities were demonstrated	Signature of accreditor and date
Determination		
Ability to communicate – written		
– oral		
Punctuality		
Attendance		
Reliability		
Concern for others		
Co-operation		
Ability to work alone		
Initiative		
Self-confidence		
Sociability		
Presentation of work		
Ability to discuss		
Adaptability		
General behaviour		
Maturity		

Other comments the student wishes to make:

_____

_____

Students must refer to separate sheet describing above qualities.

Figure 21 Negotiated record sheet

numerous and beyond the scope of this text), but merely an indication of the main types of record that are influencing the development of profiling.

Differences in opinion still suggest that decisions have to be reached about the process and product of profiling. Choices need to be made between a norm-referenced grid approach, a pre-specified criterion checklist, or a free comment format. The kind of evidence to be used must also be decided upon – for example, formal tests, traditional marking, teacher observation or pupil self-assessment. The content has to be decided on – which skills, which achievements, which dispositions, which feelings? Who is going to record – is it to be the student alone, the teacher alone, or both together? Will 'all' the school curriculum be assessed? Finally, will profiling be centrally controlled through the DES or LEAs, or will individual institutions have the freedom to choose? Further implications of the methods of profiling discussed above will be dealt with in Chapter 5.

I believe that a trial-and-error period has been undertaken, where experience will highlight what works and what is to be selected and used in the future as good practice, although much also depends on decisions taken by the DES.

Issues associated with profiling are common to all subjects and science is no exception. What may be different are the knowledge, skills, concepts and attitudes involved in science, but even here there is a large overlap with other disciplines.

PROFILING IN SCIENCE

In looking at assessment issues in Chapter 3, inherent methods of recording for nationally important schemes were discussed. This section, therefore, cannot be read in isolation from what was discussed in the last chapter.

In science, process skills have been important in establishing criteria for referencing. Hilton (1984) provides a checklist of process skills (used by ILEA for Further and Higher education) that was derived by teachers from the skills of practising science:

The student has demonstrated the ability to:

1. Use observation to identify similarity, difference or change.
2. Use common criteria to sort or classify as in classification systems.

3. Estimate and approximate within the range of accuracy appropriate to the task.
4. Measure in appropriate units.
5. Calculate using appropriate units.
6. Draw, interpret and use representational diagrams, charts, symbols and formulae (including, for example, graphs, tables and histograms).
7. Read reference material to extract content for a given purpose.
8. Identify and describe the nature of simple problems (recognise and describe simple environmental, social and practical problems).
9. Suggest explanations, causes or solutions (by applying knowledge and experience in a new situation, e.g. to form a hypothesis or use models and analogies to explain ideas).
10. Plan investigations following instructions or own plan (could refer to designing experiments and using the idea of a control).
11. Assemble components for an investigation or task (assemble equipment, apparatus, living and non-living material into a working whole).
12. Test explanations and monitor outcomes (by investigation or operating procedures and monitoring outcomes).
13. Record observations and procedures in suitable form (using diagrams, tables, graphs, notes or reports).
14. Draw and explain conclusions at an appropriate stage (by making logical deductions and assessing their implications).
15. Evaluate investigation or procedure (including, for example, suggesting modifications or improvements to procedure if appropriate).
16. Assess own performance.
17. Use and maintain tools, equipment and materials.
18. Use computers for information retrieval or learning.
19. Care for living things.
20. Work independently on a given task (in an effective systematic way).
21. Co-operate with others on a given task (in an effective systematic way).
22. Practise health and safety procedures.
23. Perform tasks within the constraints of available time and resources.

The checklist includes skills readily identifiable as 'scientific' (for example, 1, 2, 8, 9, 10, 11, 12, 13, 14, 15, etc.) and others that are necessary if performance is to be communicated (for example, 4, 5, 6, 7, etc.) and achieved co-operatively (for example, 21, 22, 23). These skills are regarded as context-free and can therefore be used with any syllabus.

Others have produced similar lists. Hodson and Brewster (1985) group processes into three categories: planning investigations, carrying out investigations and interpreting and learning from investigations (see Appendix 6 for complete list). These authors agree that the list is not an 'exhaustive and comprehensive collection of all possible scientific activities'. It is also important to note that not all scientific investigations will comprise all the processes listed in Appendix 6. Harlen (1983), working with middle school children, divides processes into: enquiry skills, scientific attitudes and concepts. Often she sub-divides processes into 'earlier and later phases' and 'later phases' (the former phase represents children up to 9 years and less mature older pupils and the latter phase up to 13 or 14 years or more mature pupils). Appendix 7 indicates the full list of processes. Another similar list is that devised by the APU, for 11-, 13- and 15-year-olds.

Figure 22 illustrates an example of ILEA's Further and Higher Education (FHE) Curriculum Development Project Process Skills Record Sheet. Evidence of students mastering a skill, as demonstrated in practical work, can be recorded on the sheet to provide a profile of the student's ability in science.

Profiling may describe more fully students' competencies than is possible in examinations because a wider range of abilities, concepts and attitudes can be assessed and recorded. Process skills assessed can also be extended across other curriculum areas to relate the learning in other subjects. Profiling can also allow for student–teacher negotiation (this would not be possible, say, in a practical examination), which provides the benefits previously mentioned.

It is appropriate that this chapter should end with a consideration of how profiling might operate through an 11–16 school for a typical pupil. Profiling practice is so diverse that whole texts have been written on the subject; here, only an outline of a notional scheme is intended.

SKILLS / TASK & DATES	Ink	Seedlings	Picture frame
P1 Using observation	✓	✓	
P2 Using common criteria to sort or classify	✓		
P3 Estimating and approximating	✓	✓	
P4 Measuring in appropriate units	✓		
P5 Calculating using appropriate units			✓
P6 Drawing and using representational diagrams, charts, symbols and formulae	✓		✓
P7 Reading reference material to extract content for a given purpose	✓	✓	✓
P8 Identifying and describing simple problems			✓
P9 Suggesting explanations, causes or solutions	✓		✓
P10 Planning investigations following instructions or own plan		✓	✓
P11 Assembling components for an investigation or task	✓	✓	✓
P12 Testing explanations and monitoring outcomes			
P13 Recording observations and procedures in suitable form	✓		
P14 Drawing conclusions at an appropriate stage	✓	✓	
P15 Evaluating investigation or procedure	✓	✓	
P16 Assessing own performance	✓	✓	✓
P17 Using and maintaining tools, equipment and materials		✓	✓
P18 Using computers for information retrieval or learning			
P19 Caring for living things		✓	
P20 Working independently on a given task	✓		✓
P21 Co-operating with others on a given task		✓	
P22 Practising health and safety procedures		✓	✓
P23 Performing tasks within the constraints available time and resources		✓	

Source: Hilton, 1984. (Reproduced with permission.)   ILEA © 1983

Figure 22   ILEA FHE Curriculum Development Project Process Skills record Sheet

Source: Hilton, 1984. (Reproduced with permission.)

PROFILING IN ACTION

The model to be described is based on four areas:

  subject assessment
  cross-curricular skills assessment
  personal and social skills assessment
  achievements and experiences.

All students of all ability ranges participate in such a scheme from entry into the first form until the end of the fifth form when the summative profile is issued to them.

Two compatible profiles exist in this scheme, one formative to motivate the student and assist in the learning process as well as to provide information for teachers and parents, the other summative on completion of compulsory education. The summative profile will inform the student of his or her achievements and could be useful to parents and potential employers.

Where there is student–teacher negotiation at the formative stage, the teacher has to sign the mutually agreed mark, grade, categorisation or comment to show agreement with the student. The reason for this is twofold. First, the student needs to be trained in self-assessment and, secondly, the profile becomes a more valid document for potential employers. The formative assessment is a continuous process, whereby the teacher records and negotiates evidence as it becomes apparent; this is particularly the case with affective skills and achievements and experiences. With subject assessment, and the assessment of cross-curricular skills, the teacher may need to 'contrive' situations in order to ensure that the scene is set for the abilities to be demonstrated. Since the main aim of the profile is to help students, only positive statements are made; nevertheless, areas that require improvement will be identified during the negotiation and the teacher can decide on remedial action.

For each subject, departmental decisions are made as to the formative nature of the profile and precisely what is assessed within that subject. Earlier in this text, for example, some indication was given as to how practical science abilities may be recorded. Practical science lends itself to the use of detailed checklists, where a tick in a box may indicate the demonstration of a skill by a student (see figure 22). In English, however, a more subjective format involving comments may be favoured.

Evidence for recording is obtained from observations and discussions during lessons and also from written reports. When the teacher has 'jotted' down the assessment, it needs to be discussed before it is finalised. Subject teachers have to organise their lessons to ensure that negotiation can take place during teaching time, but this must be unobtrusive to the teaching and learning process. Departmental decisions have to be made here about the strategy to be employed. Team teaching is one way to free a teacher in order to negotiate with students. Teachers can also use form-period time to continue this work. Every teacher of a particular child is expected to participate in the profiling and contribute to the record of achievement.

Cross-curricular skills are demonstrated in more than one subject area, although each skill may not be assessed in every subject area. These are skills concerned with learning in the cognitive domain, organisational skills and skills of motor development. The classification of each skill area is as follows:

Cognitive Skills	*Organisational Skills*	*Motor Skills*
Learning skills	Following instructions	Writing
Listening	Dealing with information	Neatness
Reading	Use of graphs and tables	Use of equipment
Oral skills	Planning	Measurement
Problem-solving	Evaluation	Physical co-ordination
Memory	Concentration	Manual dexterity.
Calculations		
Use of numbers		
Visual understanding		
Creativity		

Care is taken to avoid duplication of information on the profile. Departments are expected, for example, to avoid the recording of information noted on the cross-curricular section of the profile.

After student–teacher negotiation, comments from a comment bank are submitted to the year tutor. The 'better' marks are collated by the year tutor to produce a profile of cross-curricular skills.

The recording of personal and social skills is carried out in order to assist the student and must be done with sympathy and sensitivity. The written comment approach is adopted and the wording is negotiated with the pupil. The assessment should be based upon specific evidence and not on an impression. In this scheme, all teachers familiar with an individual can contribute to the profile at such times as evidence of certain attitudes is demonstrated by the pupil. The teacher merely recalls the event as it happens; it is up to the reader to make the assessment. Year tutors will collate the responses for each student under the following headings:

everyday coping skills	responsibility	initiative
dealing with people	ability to work in a team	resourcefulness
speaking	punctuality	appearance
sociability	attendance	attitude
working with those in	self-awareness	enthusiasm
authority	self-confidence	reliability
sensitivity	flexibility	perseverance.
leadership		

Written comments are adopted for achievements and experiences and appear under separate headings. For experiences, students merely make a free subjective response listing their own experiences. The achievements of students are also listed in a subjective manner, but in this case much more negotiation occurs, since students may not always appreciate their own achievements. Students can ask to record achievements and experiences at any time, but the opportunity should be made available, at least once a term by the form teacher during form periods. Recordings are submitted to the year tutor for collation.

Profiling is an unbroken process – the profile is continually added to by teachers submitting assessments to the year tutor. At the end of each year, copies of individual records covering the four areas are bound into a booklet and given to the pupil. All the collating and compiling is undertaken by the year tutor. Only the assessments submitted to the year tutor by a given point in the summer term are used for the end-of-year summative profile. The profile is added to over the five years, culminating in a detailed summative profile which is given to all students at the end of their compulsory education. Throughout this period pupils and parents (through staff) and staff have access to the profile at any time. I believe that if the system is to be a success it should be the central focus of the school reporting system.

Much of the time required for profiling will come from the removal of the traditional reporting and recording system. Also, the use of computers and word processors should make the task easier and less demanding of teacher time.

Chapter 5

Implications of Profiling

Many teachers are enthusiastic about the development of profiles. In a survey conducted by Hitchcock (Mortimore, 1984) teachers were eager to develop a profile system. Many reasons were given but the following three were mentioned by staff in all the institutions:

- dissatisfaction with the current examination system, and its effect upon so many of the schools' population
- determination to provide young people with a worthwhile statement relating to their achievements
- a wish to motivate pupils by involving them in a form of assessment more relevant to their interests and needs.

Teachers also viewed profiles as a way of providing comprehensive information about their pupils and of recognising pupils' strengths and aiding the diagnosis of weaknesses. In the opinion of these teachers, this type of recording can aid staff development in assessment skills. They also expressed an awareness of the curriculum implications of integrated learning and assessment procedures.

For clarity and ease of presentation, the implications of profiling will be discussed under separate headings, but it must be borne in mind that the arguments are interrelated. One cannot discuss examinations in total isolation from the rest of the curriculum or pupil–teacher relationships or the effects on the institution, etc.

IMPLICATIONS OF PROFILING FOR THE EXAMINATION SYSTEM

The examinations system is an inadequate form of assessment. One of the major reasons given by the NUT for its support of profiles is the inadequacy of the examinations system to test the wider range of 'skills, abilities and qualities which secondary schools seek to develop in their pupils'. Examinations, particularly the traditional, most widely used, end-of-course written paper, can assess only a narrow range of skills or abilities – many are designed primarily to test knowledge acquisition (cognitive skills). It is recognised that almost all our present assessment procedures have been directed towards such cognitive outcomes. Our competence in measuring the skills of the cognitive domain far exceed our ability to measure attitudes or other affective skills. In a document from Her Majesty's Inspectorate (HMI), 'Curriculum 11–16' (1970), it was stated that examination results offer some evidence of genuine application to work, but little is discovered about other educationally desirable qualities.

There is some general criticism of the inscriptive techniques that are used by examination boards to discover students' abilities. It is believed (Rowntree, 1977) that the written product used to assess students is not going to necessarily reveal qualities of creativity, fluency, imagination, reasoning, drive and persistence. In fact, as far as creativity and imagination are concerned, these could be a positive liability to success under the present examination marking system, where consistency is of prime importance.

In science education, assessment by examination alone is limiting. Progress in science should also be assessed. Assessment should 'recognise the importance of skills and processes of science as well as rewarding the ability to reproduce and apply scientific knowledge' (DES, 1985a). Assessments made should allow pupils to demonstrate what they know, understand and can do rather than what they cannot do.

The Schools Council (1981) argues that owing to the limitations of other forms of recording achievement, secondary schools ought to develop their capacity to write detailed profiles of their pupils. Concern that examination certification does not adequately express pupils' developments or attainments has been a major reason for the development of pupil profiles. Stevenson (1983) stresses, however, the fact that 'Criticisms of one form of assessment . . . may not be regarded in themselves as justifications for another form of assessment'.

Concern about the effects of examinations is far more wide-reaching. Burgess and Adams (1980) claim that the examination system unduly influences and controls the curriculum in secondary schools, their internal organisation, management and discipline.

The next section will discuss alternative methods to the traditional examination system that have been adopted by various courses for assessment and certification.

IMPLICATIONS OF PROFILING FOR THE CURRICULUM

The word curriculum will often be used in this section in its narrower sense to refer to subjects and courses.

Curriculum and assessment objectives should be planned, developed and evaluated jointly. Garforth and Macintosh (1986) believe that such a strategy must be flexible enough to respond to national curriculum and assessment initiatives and also to local and institutional needs. This statement may be at odds with the view of the FEU of the DES (1982) that 'for a common curricular framework, the format of the profile should be common'. Here the profile is seen as a comparable, though not a competitive, document.

Garforth is heading a DES-supported profiling pilot scheme that is endeavouring to produce profiles that reflect the individuality of each educational establishment involved in the scheme. It is anticipated that such an approach will unify the total educational experience for everyone concerned. It is true to say, though, that it is very demanding of teacher time.

Sometimes, the learning experience clearly identifies the assessment objectives, but, sometimes, curriculum opportunities need to be designed so that specific skills, attitudes or concepts can be assessed. Assessment can also assist the curriculum in the evaluative process. In order to evaluate the curriculum a structured assessment programme is needed that selects the assessment techniques most appropriate to the achievement of a particular aim. An assessment programme is also important for teachers to evaluate the success of their teaching, by measuring the students' achievement of specific aims and objectives. This process may be instructive as to the suitability of the content, the pace of the work, the degree of difficulty, the teaching methods used and the materials.

A basic principle of student profiles is that they are recording a wide and diverse range of assessments of knowledge, skills and experiences. These assessments should not necessarily be amenable to reductionist techniques, such as grading.

The new profiling and assessment spirit has influenced many of the courses now operating. The new GCSE science syllabuses, for example, are allocating at least 20 per cent of the course marks to school-based assessment of practical work. This spirit of assessment has also affected the traditional A-level science courses of which many are now including elements of school-based assessment. Disappointingly, though, in the end, students merely end up with a grade to reflect their abilities and achievements in a subject.

There has also been a wide uptake of the 'new' assessment techniques and profiling by the pre-vocational courses. The Technical and Vocational Education Initiative (TVEI) was launched in 1982 to stimulate the provision of technical and vocational education for 14- to 18-year-olds from a wide range of abilities. The final TVEI certificate includes a pupil profile accrediting skills and qualities not mentioned elsewhere. Some of the LEAs involved are even seeking separate accreditation for new courses from the examining boards. The Certificate of Pre-Vocational Education (CPVE) is a one-year full-time course that replaces the Certificate of Extended Education. Some students studying for TVEI leave at 16 and go on to CPVE. The CPVE is designed to give young people the basic skills, attitudes, knowledge and social competencies they will need for success in adult life, including work. This course caters for the 'new sixth', the 100,000 or so pupils who wish to stay on at school or take a full-time college course but do not want to take A-levels or a purely vocational course. The course is designed for a wide ability range and the DES hopes it will provide a route back into A-levels or on to Business and Technician Education Council courses (BTEC). CPVE is externally moderated and based on attaining specific skills (criterion-referenced) rather than on peer group standards (norm-referencing). Formative assessments will help students to monitor their own progress and ensure that the programme is tailored to individual needs. Summative assessments provide the basis for the final report of students' achievements that will be used by employers or admission tutors. Everyone who completes the course will receive a certificate with a profile of experience and achievement during the course.

The BTEC General Award in Business Studies was introduced in its present form in 1984 and aims to provide a broad general education for those 16- or 17-year-olds with few academic qualifications who are interested in a clerical or sales career in industry, commerce, distribution and the public sector. BTEC places a premium on learning skills through vocational experience. Teachers set and grade practical assignments to assess care areas and BTEC is responsible for moderation. There is also an external examination for the core modules at the end of the course. Although school- or college-based assessment takes place, there is no profiling involved.

The Royal Society of Arts (RSA) and City and Guilds of London Institute (CGLI) (see Chapter 4) have both produced pre-vocational profiles. The RSA offers as four main vocational preparation courses: distribution, clerical, basic clerical procedures and basic receptionist/telephonist skills. These were first offered nationally in 1977. All courses share a common core: communication skills, numeracy and career and personal development. The Vocational Preparation course in basic receptionist/telephonist skills is based partly on an examination. The RSA also runs the Practical Profiles scheme. It provides assessment for basic skills that can be applied to different vocational contexts, for instance catering or engineering. Schools and colleges can either use the Practical Profiles to accredit existing courses or devise courses to satisfy the Practical Profiles criteria. The main areas are communication skills, numeracy skills, process skills, computer literacy and information technology and a profile of general skills such as gathering information and identifying and solving problems. All courses involve profile certification.

The CGLI Vocational Preparation (general) 365 course is the forerunner of CPVE and was designed to give students from 14 upwards a chance to sample a wide range of vocational activities. It was first introduced as a pilot scheme in 1981–82. Schools and colleges are free to design their own courses, which must meet the aims of the scheme and follow its basic curriculum structure. These courses are aimed at a wide range of ability but the average student is of CSE grade 3 standard. The basic core includes: communication skills and numeracy; economic, social and environmental studies; vocationally-orientated studies; extension studies, which encourage students to develop their personal interests; guidance education; process skills such as literacy, numeracy, social skills and computer literacy, which are learnt in a variety of different situations; and work experience, which is not compulsory.

The CGLI sets tests for communication skills and literacy. Each student also has a Profile Report Form, which records the basic abilities and experience that he or she has gained throughout the programme. The Institute moderates the profiles compiled by teachers.

The above examples show wide uptake of both internal assessment and also of profiling. It needs to be pointed out, though, that many subject teachers are nevertheless reluctant to create courses that could be divided into easily recognised, assessed and recorded individual components, because they feel that such itemisation might undermine broader curricular objectives such as concept formation or appreciation of literature or other arts. Indeed, it has been further suggested that the specification of basic skills might lead to a narrower, more rigid curriculum rather than the broader, more responsive one which it is often claimed would result from profiling. If this were to happen, profile reports would have a similar effect on the curriculum to that of public examinations, as previously discussed. However, such dangers should be avoided if teachers continue to be guided by the aims and objectives of the course, of their own subject department and of the school as a whole. These intentions should encompass both the narrow specific objectives and the broad aims, which should determine and permeate the educational experiences of the students.

IMPLICATIONS OF PROFILING FOR RECORDING AND REPORTING SYSTEMS

If a profiling system is to be introduced in schools then it must form the basis of the recording and reporting system with regard to students. A carefully designed profile would provide parents with far more detailed information than they have traditionally received about progress, attainment, achievements and specific learning difficulties via the traditional school report. It would be absurd and excessively time consuming to operate a profiling system in addition to traditional recording and reporting.

In advising schools and teachers on setting up a profiling system, Garforth and Macintosh (1986) state: 'you may wish to maintain for each student a confidential file containing the minimum of information'. Schools are frequently required to provide employers or further education institutions with confidential references for their pupils. Some schools surveyed by Balogh (1982) expressed the hope that profiles would eliminate the need for confidential references to be prepared, although many employers insist on the completion of their own forms for this purpose.

It is also important to consider the question of confidentiality in a wider context. Under the Data Protection Act 1987 anyone can request information on themselves that is retained in computer systems. It does not appear likely that schools will be exempted from such legal requirements. Although there are strong educational reasons for not having confidential reports, legal requirements will ensure, unwittingly, that the system remains 'open', in the true spirit of profiling.

There is now great support for the idea that every pupil completing a course of secondary education should be offered a balanced documentary record to that effect, incorporating a range of information about the pupil and his or her experiences of schooling. Reservations have been expressed, though, that such records should not take the form of extended testimonials, as these are particularly prone to subjective bias. Desirably a document could contain evidence of a reasonable level of attainment in language and mathematics, the number of courses a pupil has followed, with some indication of the standards achieved, the pupil's extracurricular interests and contribution to the life of the school.

Employers will be one of the main customers of any new form of certification such as profiles. Balogh's (1982) findings indicate a variety of responses from this group. It was felt by some that profile reports would only be useful if they offered 'more than a traditional testimonial's positive statements'. The idea of portraying the negative as well as the positive attributes of the students is frowned upon by many teachers. This is because some students who do not succeed in school may reach high levels of success outside. Some employers will use the telephone to find out about employees, or even take up references after the employee has commenced work. Again, teachers disapprove but do little about it, for fear of prejudicing students' employment chances. Employers seek easily assimilable references, so some prefer their own documentation, which is familiar. It has been indicated by Burgess and Adams (1980) that employers are only prepared to devote a maximum of 15 minutes to considering an application, and that they place greater emphasis on interviews than on the profile.

There is evidence that many companies do not rely heavily on the results of external examinations to fill vacancies. Many jobs are, in fact, unconditionally filled before examination results are available, despite stated examination requirements for the job. A great number of companies are more interested in the qualities of individuals, such as common sense, conduct and punctuality. This is a strong argument for the use of profiles.

Some may argue that a grade is less amenable to misinterpretation than a profile and the provision of a profile by a school involves the school in complicity in the process of job selection. This may be antagonistic to the idea that profiles minimise the selective and discriminating functions of public examinations. At least, with a profile a wide picture of the student's abilities is built up. The reductionist grade tells us little about the individual concepts, attitudes, knowledge or abilities a student may possess.

There appears to be even less co-operation between further education colleges and schools. This is a reflection of the rivalry that has often taken place for sixth formers – a conflict that has often resulted in a college not accepting a school's profile.

Stevenson (1983) believes that 'profiles as an alternative to public examinations are unlikely to be seen as credible by either parent or employers'. I cannot accept such a pessimistic viewpoint. Enthusiasm from pupils and the more relevant information made available in profiles will make them a readily acceptable document to parents. It does appear, though, that before there is a wide acceptance of the profile, in its summative form as a reference, much needs to be done, both in educating employers and in making sure that our arguments for the format and content of the profiles are sound.

IMPLICATIONS OF PROFILING FOR THE PUPIL

Although there is great concern about certain ambiguities in profiling, many that broadly sympathise with it do so because of the potential implications for pupils. These implications include: the enhancement of pupils' self-esteem and self-awareness; the recognition of the whole range of pupils' achievements; and the improvement of teacher–pupil relationships.

One of the dominant reasons for developing records of

achievement is to enhance pupils' motivation. In Balogh's (1982) study and in the follow-up study of Goacher (1983) it was found that many schools profiling did so because they felt pupils would benefit from improved motivation. Hargreaves (1986) points out that there has been a weakening in the incentive to work, to aspire and to conform in modern Britain. He cites unemployment and the poor prospects of work as major reasons for this. The belief that qualifications are the key to a successful future and a well-paid job is no longer valid. Habermas (1976) sees the problems of motivation in school as one more face of a multifaceted problem of loyalty and commitment to the present social order in society as a whole.

Earlier, the problems associated with employers accepting the profile were discussed. Hargreaves proposes that the two purposes, motivational and selective, are fundamentally incompatible. The move to involve employers has partly arisen from the desire to attain wider acceptability for the profiles and also from the dissatisfaction of many employers with the examination results and traditional reference as a basis for selecting.

Employers appear to want easily scannable, summative records of achievement, with numerical grades, ticked boxes or blocked charts; but crude grades do little to enhance pupil motivation, especially for those pupils who consistently score low marks. Aware of the effects of such methods on student motivation, the DES advised strongly against systems using 'ticks in boxes or numbers or letter gradings' or references 'to failures or defects' (DES, 1984). The DES suggested that one way of overcoming this problem would be to produce a succinct prose summary involving 'sentences written for each pupil'.

For motivational reasons, it is important that the pupil contributes to such a record. A more closed type of prose summary, such as the comment bank, is favoured by many schools, because of the low cost involved and its efficiency (discussed in the previous chapter). In such a situation, however, student involvement is reduced to making selections from lists of statements that are not negotiable. The effects of such methods on motivation are less positive than are those with a truly negotiated profile.

There are also fundamental differences in the envisaged content of records. Activities and experiences that often signal employability are loyalty, obedience and conformity, with little weight placed on initiative. The IPM on the other hand (see Chapter 4) has produced a list of what employers seek that more closely matches similar lists for intended educational outcomes. Nevertheless, there is the dilemma that improvement in selection orientated records could bring about a weakening of motivation.

Another area of great concern for records of achievement is their potential use as surveillance mechanisms rather than as devices to encourage independence. It has been argued by Hargreaves in *Profiles and Records of Achievement* (Broadfoot, 1986) that the personal record component of records of achievement,

> with its assessment and monitoring of affect as well as intellect, of personality as well as performance, according to a carefully graded schedule of systematised review, is in fact bound up with a more generalised trend towards the development and implementation of increasingly sophisticated techniques of social surveillance within society at large.

This has been an argument used in favour of the Personal Record of Achievement where there are no hierarchical elements of partnership and negotiation and where pupils own and control the summative statement.

Undoubtedly, profiles can be used to motivate and encourage independence among our pupils, but pitfalls in format and procedures to be adopted must be avoided at the developmental stage. Profiles also enable the monitoring of progress on a continuous and flexible basis and permit speedy and well-informed remedial action to be taken. Support for such an approach is reiterated in two recent DES reports.

The Science Working Group was set up by the Secretary of State for Education and Science in July 1987. Its terms of reference were to recommend attainment targets and programmes of study for science and technology within the framework of a proposed national foundation curriculum. The group's interim report (DES, 1987) recommended assessment procedures that are widely regarded by many teachers as good assessment strategies for profiling. The report proposes that assessment should assist in the pupil's learning process and that the wide range of knowledge and understanding, skills and attitudes should be incorporated in the assessment scheme. It also supports the view that assessment forms an 'integral and continuous part of the learning process'. Importantly, it recommends 'that continuous assessment leading to profile reporting should be a pivotal aspect of the overall provision'. This working group is against the idea of merely reporting at the ages of 7, 11, 14 and 16.

The above recommendations are also voiced in the report by the National Curriculum Task Group on Assessment and Testing (DES, 1988). The latter report makes the teacher the central figure in the assessment procedure and proposes that tests should only form part of the assessment. If this report is rejected and pass/fail tests at the ages of 7, 11, 14 and 16 are implemented, there is the danger that the aforementioned tests will become the focus of the educational process, especially if the marks of the tests are published to enable parents to choose between schools.

It is important that such tests are not elevated to a position where they dictate the style and content of teaching to the extent that the real needs of the student are not met. It must be borne in mind that profiling saw its origin with dedicated teachers who wanted to enrich the educational process for their students. Formal pass/fail tests would act against much of what teachers have endeavoured to achieve through profiling.

IMPLICATIONS OF PROFILING FOR THE TEACHER

The attitude of teachers is one of the most important factors in determining success or failure in the introduction of profiles. There is general agreement among teachers involved in profiling that the completion of profiles, and the necessary consultation with other staff involved, is extremely time consuming. In the Schools Council survey by Balogh (1982) some of the teachers questioned resented the extra work, which they saw as interfering with more important activities. Teachers who were committed to the idea of profiling, on the other hand, were not resentful of the extra time and work involved. At a time when teaching morale is low and there is general disagreement between the Government and teachers over pay and conditions of service many resent the extra demands that are being made upon them. Profiling is seen by a large number of teachers as part of the global problem associated with conditions of service.

Commitment is weakest when a profile scheme is introduced without proper consultation between senior teachers and other staff. It is evident that for any profile scheme (as for any new major initiative in a school) to be successful teachers must be fully committed beforehand and convinced of the need for and value of such a scheme.

In spite of what was discussed earlier, it appears that the degree to which a profile would be valued by employers and other users is a major factor in convincing both teachers and pupils how worthwhile an innovation a profile scheme is to them. If the profile is only truly going to be regarded as worthwhile by pupils and teachers if it is valued by employers, this in itself may be a strong argument in ensuring that the differences of opinion between employers and schools are ironed out and that employers are educated into the spirit of profiling.

Two out of three teachers taking part in the Schools Council project directed by Goacher (1983) found it difficult to complete profiles satisfactorily. Traditionally, teachers' expertise lay in assessing attainment in subject-based skills and they have had little or no training in the assessment of cross-curricular skills or personal qualities. The completion of a profile covering the wide range of skills and qualities identified in the Schools Council criteria would present many teachers with tasks they feel unable to perform adequately. This has significant implications for the in-service training (INSET) of teachers. Furthermore, the introduction of formative profiles would require many teachers to rethink their aims, objectives and teaching strategies.

As already mentioned, it is important to regard assessment as an integral part of curriculum development. Goacher acknowledges that profiling is a good way of improving teaching, as it allows teachers to take greater responsibility for assessment. An attainment profile can, in fact, act as a useful diagnostic tool for teachers in monitoring their own teaching.

Obviously, traditional perceptions of the role of the teacher will be greatly challenged by a system that regards students as partners in a negotiated programme. In some institutions, the change is quite radical; the move is made from a situation where the teacher decides to a phase when decisions are negotiated between student and teacher and finally to a programme in which the assessment decisions are made by the pupil. The student–teacher relationships are undoubtedly going to change when a negotiated or student recorded profile programme is undertaken. What many in education hope for, though, is that such relationships will be enriched, providing mutual trust and respect.

Garforth (1986) believes that, 'the development of a profile scheme can provide cohesion and mutual support by integrating the curriculum, the pastoral structure and the recording and reporting systeem'. It is inevitable, therefore, that the relationships between members of staff will also change.

IMPLICATIONS OF PROFILING FOR LEAs

Profiling requires additional resources for schools. These include adequate material resources such as paper, reprographic equipment, etc. and clerical and ancillary support.

The skills of staff will need to be redistributed and extra non-teaching time will need to be made available and, by implication, extra teaching staff will need to be employed. As previously mentioned, INSET programmes for participating staff will be needed. Much of the training will have to be school-based or school-focused in order to meet the needs of each scheme.

There is a large pool of potential support and information for teachers and this must be drawn upon. It includes expertise of further and higher education and also that of advisory staff and inspectors. Periods of secondment will also enable teachers to develop the necessary skills. It is up to the LEAs to enable schools to undertake such a programme of training in order to meet the new demands posed by profiling. Most of all though, LEAs need to participate in genuine consultation with teachers in a spirit of flexibility and goodwill on both sides.

CONCLUDING COMMENTS

Clearly, science is only one part of the total enterprise to which assessment and profiling must refer.

The major implication of profiling for science education in schools is that it will demand a greater degree of internal assessment. This will enable the assessment to reflect the essentially practical nature of science and will encourage the development of its skills and processes. All this must be seen in the context of the principal reasons for profiling, namely: to enhance pupils' self-esteem and self-awareness; the recognition of the whole range of achievements; and the improvement of pupil–teacher relationships.

Appendix 1

GCE O-Level Practical Examination Paper*

Additional Information not supplied to students:
D 31   = potato
D 32   = potato exposed to air
D 33   = sugar solution
D 34   = half of apple
O-Level examination paper:

BIOLOGY
Practical Test
(One hour)

Answer both questions.
Written answers should be kept to the lines.
Drawings should be made in the spaces provided.
Use sharp pencils for your drawings.
Coloured pencils or crayons should not be used.
No additional sheets of writing paper to be inserted in this book.
Work on additional sheets will not be marked.

1. Begin this question as soon as possible.
    Place one of the pieces of D 31 against a ruler and cut it squarely at both ends to a length of 6 cm. Place this in the test tube containing solution D 33.
    Repeat the operation with the other piece of D 31, placing this piece in the test tube of water.
    Record the time here ....
    Leave the two test tubes for at least 20 minutes. Meanwhile begin work on question 2.
    After at least 20 minutes, remove the piece of tissue from the test tubes and measure the length of each of them. Record these measurements in the table below.

* From the University of Cambridge Local Examination Syndicate. (Reproduced with permission.)

	First measurement	Second measurement	Difference in length
Piece in D 33	6.0 cm		
Piece in water	6.0 cm		

(a) Record differences in texture and appearance between the two pieces

....

(b) (i) What similarities are there between the piece that was in D 33, and the piece D 32 (which has been left exposed to air after cutting)?

....
....

(ii) What differences are there between the two pieces?

....
....

(c) How do you account for the result with D 31 in water?

....
....

(d) (i) Account for your observations and measurements of D 31 (in solution D 33) with reference to the cells in the tissue.

....
....

(ii) What would cells in this piece of D 31 have in common with cells of D 32?

....
....

(iii) What can you deduce about solution D 33?

....
....

2(a) Make a clean transverse cut through the middle of D 34. Cover one of the newly exposed surfaces with iodine solution and leave it while you make a large, detailed drawing of the other, untreated, surface in the space below. Labels are not required. You should examine the specimen with a hand lens and it might be helpful after a few minutes to make a fresh cut.

Measure accurately and record the widest distance across D 34. Use this information to calculate the magnification of your drawing.

Distance across D 34 ....   Magnification ....

With the help of a small, simple sketch (not another detailed drawing) show the effect of the iodine solution on D 34.

Describe the distribution of the staining (colouration) and say what this indiciates.

....

....

....

(b) Take a piece (about 1 cm^3 or just enough to slide easily into the test tube) of the main fleshy part of the half of D 34 which was not treated with iodine solution and place it in a test tube one-third filled with water. Place your thumb over the end of the test tube and shake the tube vigorously, to mix the contents, for about 20 seconds.

(i) Describe the appearance of the contents after shaking.

....

(ii) Add Benedict's or Fehling's solution and warm the tube slowly.

Observations ....

....

Deductions...

....

(c) How do these tests help you to understand the usefulness to man of D 34?

....

....

....

Appendix 2

Nuffield A-Level Biology Practical Assessment*

Assessment of practical work

The instructions for practical work given below must be used for the 1984 and subsequent examinations. Revised 'Instructions and guidance for teachers' were circulated to all centres in September 1981. Additional copies can be obtained from the Secretary, The Joint Matriculation Board, Manchester, M15 6EU.

The assessment of the practical course work is an essential part of the examination. It is concerned with two aspects of the syllabus. The assessment is carried out by the candidates' teachers. Instructions and forms for assessment must be obtained at the start of the first year of the course from the Secretary, The Joint Matriculation Board, Manchester, M15 6EU.

1. *Assessment by the teacher of practical course work*

The practical work in the syllabus is mainly concerned with providing evidence of biological principles from experiments. It is also used to provide information on which an assessment of practical abilities is based. Teachers may employ various methods to obtain this information. For example, it might be derived from

(a) exercises in the course chosen to give a representative range of practical activities (candidates should be informed when an exercise is being used for assessment),

(b) practical tests devised for assessing specific manual and intellectual skills,

(c) the practical competence of candidates recorded for extended sections of work or over a period of time.

Operational divisions of an investigation involving practical work

Assessment is based on the three divisions described below. The relative importance of these will vary in different exercises, but, overall, they should be given equal weighting.

*JMB (Reproduced with permission.)

(a) *Procedure*
This includes the assessment of manual operations and the carrying out of a correct sequence of activities in the laboratory or in the field. A reasonable overall efficiency should be sought rather than the attainment of special skills by repetitive practice.

(b) *Recording*
All methods used for recording observations and experimental results should be taken into account, e.g. short notes, sketches, diagrams, tabulated results and graphs.

(c) *Handling of results*
Assessment is made of the ability to draw conclusions from observations and data, and, where required, to propose a relevant hypothesis.

The assessment is made on the practical work of the third, fourth and fifth terms of a two-year course or on the sixth, seventh and eighth terms of a three-year course. If a student is repeating the course, the assessment is made on the first two terms of the repeated year. Assessment grades are to be sent to the Joint Matriculation Board, on the special form provided, not later than 30 April in the year of the examination. The teacher's final judgements on grading should be deferred until late in the course to allow for an adequate range of work and for evidence of improvement to be taken into account.

Assessment grading
Teachers are asked to make their assessments on a ten point scale of grades, using the following criteria for the three operational divisions.

(i) *Procedure*

10 – 9 Procedure is carried out with care, understanding and persistence. There is good appreciation of the need for, and the use of, experimental controls. Sequence of operations is well thought out and the nature of the investigation is clearly specified. Apparatus and materials are used with a high degree of skill. Capable of independent work.

8 – 7 Less understanding and persistence although instructions are followed carefully with minimum help. Recognises the need for, and the use of, experimental controls. Apparatus and materials are used effectively.

6 – 5 Instructions are followed with adequate understanding. Does not fully appreciate the need for, and the use of, experimental controls. Apparatus and materials generally handled satisfactorily but help may be required with difficult procedures.

4 – 3 Follows instructions but with limited understanding. Needs help but usually completes the investigation. Reasonable ability in the use of apparatus and materials, but help required.

2 – 1 Tends to misunderstand or ignore some instructions. Procedure is carried out with limited thought and understanding. Needs a great deal of help and has difficulty in completing the investigation. Poor or careless in the use and handling of apparatus and materials.

(ii) *Recording*

10 – 9   Record is well written and presented in the most appropriate way without help. Recordings indicate detailed and accurate observations. Data are suitably presented with effective use of diagrams, graphs and tables. Comments are included upon any shortcomings of the procedure and the limitations of the data.

8 – 7   Record is well written and presented in an appropriate way with minimum help. Recordings indicate accurate observations. Some reference to any shortcomings of the procedure and limitations of the data are included.

6 – 5   Contents and presentation of record is satisfactory but sometimes important information is omitted. Data accurately recorded but some assistance required with presentation in the form of diagrams, graphs and tables.

4 – 3   Record is incomplete with many confused and inaccurate statements. Diagrams, graphs and tables omitted or poorly presented. Impossible to analyse or interpret the data.

(iii) *Handling of results*

10 – 9   Excellent understanding of the results and of their limitations. Conclusions are supported by reasoned arguments. Aware of anomalous results. Relevant hypotheses proposed if required. Suggestions for improved procedures or further relevant investigations.

8 – 7   Good understanding of results but with less appreciation of their limitations. Sound conclusions made but fewer suggestions for improving the procedure or carrying out further relevant investigations.

6 – 5   Results are adequately interpreted and reasonable conclusions made.

4 – 3   Results not well interpreted. Conclusions are not supported by the data.

2 – 1   Misunderstandings evident in the interpretation of results. Conclusions are missing or wrongly stated.

2. *Assessment by the teacher of individual projects*

The submission of a project by each candidate is an important feature of the examination. The project should be based on a restricted topic which is selected for practical investigation, involving the design of experiments or the collection of data for analysis and evaluation. A project may develop from course work in the laboratory or the field or from special interests of the candidate. After preliminary consultation with the teacher, the project is carried out largely on the initiative of the candidate for whom it should contain some new experience such as applying a familiar technique to a new situation. Before projects are commenced teachers are advised to consult both the booklet *Projects in Biological Science* (Nuffield Advanced Science – Penguin 1970) and *The Assessment of Project Work in the Advanced Level Nuffield Biology Examination*, J. Eggleston and D. Hobson (JMB, 1977).

Each candidate is required to produce a project report of not more than 5000 words. The report, together with the assessment form, must be sent to the Moderator in accordance with the instructions that will be issued to each centre by the Board. (The instructions will give a date by which the reports have to be sent to the Moderator; this will normally be late March/ mid April in the year of the examination.)

Operational divisions for the assessment of a project

The essential activities in projects are grouped in the following five operational divisions:

   (i)   Statement of a problem;
   (ii)   Investigation of the background knowledge;
   (iii) Planning the procedure;
   (iv)   Inferences made from the investigation;
   (v)   Suggestions for further practical investigations.

The five point scale used in the assessment of a project

Each candidate's performance in each operational division is assessed by the teacher, using the following five point scale based on the average performance of the class. The criterion adopted in the use of this scale is the average performance of the class. It is not necessary to relate this to the standard of previous classes or to that of other schools.

5 – well above average
4 – above average
3 – average
2 – below average
1 – well below average

In making the assessment the teacher must avoid being influenced by the other work of the candidate.

Inside each candidate's report the teacher may include any special notes on the following points:

   (a) any project which might be penalised by the use of the operational divisions in the assessment;
   (b) any special help that might have been given to the candidate;
   (c) absences or any other difficulties which affected the project.

The teachers' grades for assessment of the projects are moderated by the Board on the basis of the candidates' reports and on any special notes which have been included in them by the teachers. Where necessary, the Moderator may make further enquiries on the project work carried out by the candidates. The moderation relates the means and spread of the assessments of the schools to one another to ensure comparability of standards.

Appendix 3

Lists of abilities to be assessed as specified by the Associated Examining Board requiring teacher assessment of practical work in advanced level biology*

Applying the Assessment Criteria

It is emphasised that each candidate must be assessed on each of the ten abilities on at least three occasions during the course of study. The same standard of assessment applies whether an exercise is carried out at the beginning or the end of the course.

Candidates must be made aware of the abilities on which they will be assessed. Having decided on an exercise, the teacher/lecturer will choose from the list of ten abilities several of which may be appropriately assessed during the course of it. Each ability should be assessed independently: for example, if a candidate displays only one of ten parts required in an exercise, "H" must be given for Ability 4. If, however, the candidate has made an accurate drawing of the one displayed part (and the drawing is a reasonably complex one), "L" must be given for Ability 5. A candidate who has made no drawing (or only a drawing from which no adequate assessment can be made) should be supplied with displayed parts and assessed for Ability 5 on the drawings made of them.

There is no limit to the help a teacher/lecturer may give a candidate provided that an "H" is awarded for all affected assessments. For example, a teacher/lecturer should assemble the apparatus if the candidate is unable to do it; the candidate must be awarded an "H" for Ability 6 for that exercise but will then have the opportunity to achieve higher assessments for Abilities 7, 8, 9 and 10 (if appropriate to the exercise).

A tick should be placed in the column headed H, J, K or L alongside the number of the exercise.

If the number of assessments carried out for one of the abilities is less than three for a specific candidate, then that candidate has failed to satisfy the requirements of the examination and will receive Grade X (no result) for the examination. The Board does not wish to impose any upper limit on

---

* (Reproduced with permission.)

the number of times that a candidate may be assessed, this being a matter entirely dependent on the centre and its resources, including the judgment of the teacher/lecturer responsible for the assessment.

Centres are reminded that all the abilities, including numbers 2, 3 and 4, may be assessed without the need for a candidate to undertake dissection of material of animal origin. Suitable exercises to test the dexterity element implicit in Abilities 2, 3 and 4 include various plant anatomical investigations and microbiological techniques. It is hoped that candidates will also handle and observe such a variety of live animals as may be available. The exercise of dissection of animal material as an investigatory technique (amongst other reasons) has its proponents, and the Board cannot either prescribe or proscribe such activity.

Ability	Grade H	Grade J	Grade K	Grade L	Exercise Number	(Tick) H J K L	Exercise Number	(Tick) H J K L	Final Grade
1 Following instructions for practical work	Unable to follow instructions supplied even with assistance (or no attempt made)	Required considerable assistance in following instructions supplied	Required some assistance in following instructions supplied	Fully able to follow instructions supplied (no assistance given)					Final Grade
2 Handling living and dead plant	More than half of material damaged (or no attempt made)	Quarter to half of material damaged	Less than a quarter of material damaged	Material not damaged					Final Grade
3 and animal material; preparing specimens and	Less than half of manipulation completed (or no attempt made)	Half to three-quarters of manipulation completed	More than three-quarters of manipulation completed	Manipulation completed to end of exercise					Final Grade
4 mounting them on slides	Less than half of parts displayed for no attempt made	Half to three-quarters of parts displayed	More than three-quarters of parts displayed	All parts displayed					Final Grade
5 Drawing	Drawing unrepresentative of specimen (or no	Material drawn from specimen with several minor	Material drawn from specimen with few or only minor	Material drawn accurately from specimen					

						Final Grade
Carrying	**6**	Unable to assemble apparatus (or no attempt made)	Major failures in assembling apparatus	Minor failures in assembling apparatus	Apparatus skilfully assembled	Final Grade
out	**7**	Few or no accurate readings (or few or no readings obtained or no attempt made)	Considerable inaccuracies in taking readings	Minor inaccuracies in taking readings	Accuracy in taking readings	Final Grade
experimental	**8**	Less than half of possible results obtained (or no attempt made)	Half to three-quarters of possible results obtained	More than three-quarters of possible results obtained	Complete range of results obtained	Final Grade
techniques	**9**	Experiment conducted in total disorder (or no attempt made)	Experiment conducted in considerable disorder	Experiment conducted in slight disorder	Experiment conducted in a well organised manner	Final Grade
Recording raw data	**10**	Raw data randomly recorded (or few or no data recorded or no data obtained)	Methodical recording of some raw data	Methodical recording of most raw data	Methodical recording of all raw data	Final Grade

Appendix 4

Midland Examining Group GCSE Assessment Details

Notes for Guidance for Teacher Assessment of Practical Work

The following notes are designed to provide teachers with information for making valid and reliable assessments of the skills and abilities required to be developed during this course.

Assumptions

It is assumed that:
1. there has been a background of practical work carried out during the first three years of secondary education.
2. thus, it is reasonable to suppose that any single assessment is a representative measure of a given candidate's practical ability, and that this is related, in part, to their previous practical experiences.

General Information
1. Each practical activity provides an opportunity for assessment.
2. These assessments should be based on the principle of 'positive achievement' i.e. candidates should be given the opportunity to demonstrate what they understand and can do.
3. It is expected that teachers will base their assessments on practical activities which are appropriate to their teaching styles and the facilities available.
4. It is anticipated that skills 1–3 will be assessed by means of teacher observation of candidates' processes and practices. It is also anticipated that skills 3–6 will be assessed by written work. Therefore all practical activities designed by the teacher should bear these points in mind. In addition, for purposes of moderation, it is essential that the written records of practical work be produced by candidates and retained by teachers until examination results are published.
5. The skills and abilities assessed must conform to the assessment objectives so that all teachers are assessing the same qualities, although the subject matter varies.

6. The standards of assessment employed are those expected at the end of the course.
7. The person carrying out the assessment should normally be the teacher responsible for teaching the candidates.
8. The assessment may be based upon written work submitted by the candidates but will also require assessment based on observation of the candidates and on discussion with them.

Recommendations

1. It is not necessary for candidates in a centre, or in a teaching group within a centre, to be assessed on exactly the same work or specific activity.
2. Assessment could be carried out on group work.
3. Where practical work involves the candidates working in groups, the teacher must ensure that the individual contribution of EACH candidate can be assessed.
4. Each practical activity does not need to encompass all the assessment skills and abilities.
5. Ideally only a few candidates within a practical group should be assessed on any one occasion where practical skills e.g. skill 2, handling apparatus and materials, are being assessed.
6. In some cases, e.g. the assessment of skill 4, recording and communicating, it may be relatively simple to assess the work of all the candidates on the basis of a written record of a SINGLE exercise.

Practical Skills and Assessment Objectives
The following skills and objectives will be assessed at THREE levels of attainment (H, high; I, intermediate; L, Low):

Skills	Assessment Objectives	Levels of Attainment
1. Following instructions	C.1.1	H, I, L
2. Handling apparatus and materials	C.1.3	H, I, L
3. Observing and measuring	C.1.4	H, I, L
4. Recording and communicating	C.1.4	H, I, L
5. Interpreting data	C.2.3	H, I, L
6. Experimental design/Problem solving	C.1.2 and C.2.1–C.2.4	H, I, L

These skills, whilst appearing to be discrete, inevitably involve a degree of overlap when it comes to assessment. For example, observing and measuring a biological structure may be assessed by means of a scale drawing. In this case, these assessment skills and objectives overlap with those of recording and communicating. Likewise assessment skills 1 and 2 may overlap to varying degrees. Whenever this happens, teachers should apportion their assessment to the two or more skills accordingly.

Awarding of Marks for Levels of Attainment
Once the candidate's level of achievement in each skill has been

identified, the marks awarded should be allocated in accordance with:

H	(9	high
	(8	intermediate
	(7	low
I	(6	high
	(5	intermediate
	(4	low
L	(3	high
	(2	intermediate
	(1	low

Frequency of Assessment
Practical Assessment Objectives 1–5
Each candidate must be assessed on a MINIMUM of two occasions for each assessment objective 1 to 5. If the teacher wishes to assess each candidate on more than two occasions, the assessment marks finally recorded, based on the three levels High, Intermediate and Low, should represent the candidate's best performances.

Each assessment objective may be assessed either on its own or in conjunction with one or more of the other assessment objectives depending on the nature of the exercise.

Practical Assessment Objective 6
This assessment deals with the candidates' abilities to perform a problem-solving exercise. ONE assessment only is required. The nature of the activity and its duration are left to the discretion of the teacher. The five sets of marks each out of 9 should be averaged to produce the final mark.

Criteria for Assessment
1. Following instructions
The ability to follow oral, written and diagrammatic instructions in order to carry out a practical investigation with due regard to safety.
The candidate will be required to follow instructions so that:
1. the apparatus is assembled correctly,
2. any procedures that are included in the instructions are followed correctly.

H	Follows instructions accurately and fluently, taking sensible initiatives when required. Safety procedures constantly and systematically incorporated into work
I	Needs occasional help in following routine instructions but ultimately is capable of completing the task safely
L	Able to follow experimental instructions given considerable guidance

For example,
Visking tubing as a model gut.
Potometer to demonstrate water uptake of cut shoot.

2. Handling apparatus and materials

The ability to use various pieces of laboratory apparatus, and biological materials efficiently, effectively and with due regard to safety in the particular contexts.

H	Handles apparatus and materials correctly and confidently. Usually recognises and corrects (without prompting) errors of assembly and safety
I	Uses individual pieces of apparatus correctly, but has some difficulties in organising equipment. Recognises and remedies errors and dangers when these are pointed out
L	Able to handle apparatus and material given considerable assistance

For example, this skill could be assessed using those examples given for skill one. In this case the candidate would be expected to carry out the investigation with such facility that a meaningful result is obtained.

3. Observing and measuring

The ability to use standard laboratory apparatus, to observe biological processes, activities and structures and to measure accurately.

H	Handles quantitative work with confidence. Uses measuring instruments correctly and relates readings to quantities. Observations are detailed and accurate
I	Qualitative observations are usually accurate in main features, but quantitative observations lack detail and discrimination
L	Able to make detailed and relevant observations and measurements given considerable assistance

For example,

Food Tests
Detailed observation of the range of results produced by a series of food tests.

Comparison of Flower Structure
Use of the hand lens to make accurate observations of flowers of different species.

Potometer
Use of graduated scale to measure rates of water uptake.

Osmosis
Investigation of changes in mass of potato chips in a series of sugar solutions.

Genetics
Scoring the results of a breeding experiment.

4. Recording and communicating
  The ability to record relevant data accurately and clearly. These data may take the form of tables, graphs, diagrams, drawings and written accounts.

H	Records experimental data accurately and clearly and in the most appropriate format. Graphs, when plotted, are correct in every respect
I	Records numerical data correctly, but units sometimes omitted or wrong. Records unsystematic and incomplete unless format given. Plots results satisfactorily onto given axes, but easily confused and misled by erroneous results
L	Able to record observations and results accurately if given a tightly prescribed format

For example,
  Osmosis
  Production of tables of results and graphs to record the results of the changes in mass of potato chips.
  Variation
  Production of a histogram to illustrate continuous variation.
  Comparison of flower structure
  Production of accurate drawings to show the differences between species.
5. Interpreting data
  The ability to interpret data and draw valid conclusions.

H	Conclusions are accurate and logical inferences from data available
I	Able to interpret data to produce logical conclusions only with considerable assistance
L	Given considerable assistance, is able to recognise a limited number of relationships within the data available

This skill can be assessed upon the conclusions recorded in the candidates' exercise books.
6. Experimental design/Problem solving
  The ability to design and carry out an investigation based on a given problem.
  (a) Identify problems and plan investigation (C.2.1)

H	Identifies problems associated with investigation without assistance. Outlines all stages of investigation accurately, clearly and without assistance
I	Some assistance required in identifying the problem but able to plan the investigation with a minimum of assistance
L	Able to appreciate the nature of the problem and plan the investigation given considerable assistance

(b) Select techniques, apparatus and materials (C.1.2)

H	Selects techniques, apparatus and materials confidently and without assistance
I	Needs occasional help in choosing techniques, apparatus and materials and in setting up
L	Able to decide upon choice of techniques, apparatus and materials given considerable assistance

(c) Organise and conduct investigation systematically (C.2.2)

H	Organises and carries out investigation systematically with due attention to detail
I	Some assistance required in conducting investigation
L	Able to organise and conduct investigation given considerable assistance

(d) Interpret and evaluate observations and experimental data (C.2.3)

H	Conclusions are accurate and logical inferences from data available
I	Able to interpret data to produce logical conclusions only with considerable assistance
L	Given considerable assistance, able to recognise a limited number of relationships within the data available

(e) Evaluate methods and suggest improvements (C.2.4)

H	Able to evaluate methods and make positive suggestions for improvements spontaneously
I	Some assistance required to evoke the need to evaluate methods and suggest improvements
L	Able to evaluate methods and make positive suggestions for improvements given considerable assistance

For example
1. Auxins are sold as rooting compounds to aid the development of roots on cuttings of Geraniums etc. Design and carry out an experiment to test the effectiveness of such a product.
2. It has been found that pieces of potatoes shrink when placed in strong salt solutions. You are provided with three salt solutions, A, B and C, and a potato. Find out which is the strongest solution.
3. When an individual plant or animal grows the rate of growth is not the same throughout its life. Design and carry out an investigation to show when a cucumber leaf passes through its greatest rate of growth.
4. The colour given by the Benedict's Test depends upon the concentration of the reducing sugar present. Design and demonstrate a method which you could use to estimate the strength of a weak glucose solution.

Although one formal assessment of this skill is being asked for, it is expected that prior to the assessment the candidate will have had experience of such an approach.

The five sets of marks each out of 9 should be averaged to produce the final mark. Fractions of marks such as 1/5 and 2/5 should be ignored, 3/5 and 4/5 should be scaled up.

(Reproduced with permission.)

Appendix 5

University of Cambridge Local Examinations Syndicate's Internal Assessment Mark Scheme for A-Level Biology

Assessment 1 – Microscopy

Skill	Mark	Maximum
(a) Construction of Hypotheses/experimental <u>design</u> Not assessed	–	–
(b) <u>Method</u> Not assessed	–	–
(c) <u>Manipulative Skills</u>		10
(d) and (e) <u>Observation and Interpretation</u>		30
TOTAL		40

Assessment 2 – Identification and Classification

	Skill	Mark	Maximum
(a)	Construction of Hypotheses/experimental design Not assessed	–	–
(b) (d)	Method Observation and results 　Observation 　Construction of a key 　Observation and use of a key 　Classification		 5 6 10 9
(c)	Manipulative skills Not assessed	–	–
(e)	Interpretation Not assessed	–	–
	TOTAL		30

Assessment 3 – Dissection

	Skill	Mark	Maximum
(a)	Construction of Hypotheses/experimental design Not assessed	–	–
(b)	Method Ability to follow instructions		7
(c)	Manipulative skills		20
(d)	Observation and results 　Accuracy of drawing 　Quality of drawing 　Labels		 5 5 3
	TOTAL		40

Assessment 4 – Experimental Work

Skill		Mark	Maximum
(a)	Construction of Hypotheses/experimental design		15
(b)	Method		10
(c)	Manipulative skills	–	–
(d)	Observation and results		15
(e)	Interpretation		20
	TOTAL		60

Assessment 5 – Fieldwork

Assessment Objectives		Mark	Maximum
(a)	Presentation		4
(b)	Aims and Background Information		5
(c)	Methods		20
(d)	Results and Analysis		20
(e) and (f)	Conclusions and Discussion		16
	TOTAL		max 60

Assessment 6 – Attitudes

This assessment should be made as late as possible in the course, preferably at the time of preparing marks for forwarding to Cambridge.

Award a mark out of 5 for each attitude separately according to the following subjective criteria:

5 = outstanding
4 = very good
3 = good
2 = satisfactory
1 = poor
0 = inadequate

Attitude	Mark	Maximum
(a) Interest/enthusiasm		5
(b) Persistence in trying to resolve difficulties		5
(c) Resourcefulness		5
(d) Co-operation in normal laboratory routine		5
(e) Respect for living organisms and the environment		5
TOTAL		25
FINAL TOTAL		255

Marks for the internal assessment of these practicals were scaled up to be in line with the other practical papers, and then weighted.

(Reproduced with permission.)

Appendix 6

Classification of Process Skills
(Hodson and Brewster, 1985)

Planning investigations
1. Identification and clarification of problems (asking appropriate questions).
2. Formulation of hypotheses.
3. Selection of suitable tests for hypotheses.
4. Design of experiments:
    (i)   analysis into component steps,
    (ii) control of variables,
    (iii) selection of appropriate procedures and apparatus.

Carrying out investigations
1. Accurate observation of objects and phenomena.
2. Selection of appropriate measuring instruments.
3. Accurate measurement.
4. Description and reporting of observations in appropriate language:
    (i) qualitative,
    (ii) quantitative.
5. Safe use of laboratory equipment.
6. Performance of routine laboratory operations.
7. Performance of specific techniques (in a profiling scheme these techniques would be listed).
8. Carrying out familiar/unfamiliar procedures in accordance with verbal/written instructions.
9. Methodical and efficient working.

Interpreting and learning from investigations
1. Processing, manipulating and organising experimental data.
2. Presentation of data in a suitable form.
3. Analysis and interpretation of data (recognising trends, sequences and patterns).
4. Extrapolation of data and generalisation.
5. Making sense of data using theoretical terms.
6. Drawing conclusions (Including the relationship between hypotheses and interpreted data).

7. Suggesting modification/improvements for further investigations.
8. Preparing and communicating an oral or written account/report in a suitable form, taking into account: (i) the content, (ii) the 'audience'.

Appendix 7

Classification of Process Skills (Harlen, 1983)

Enquiry skills
For both earlier and later phases:
- observing and exploring to observe further
- raising questions
- proposing ways to answer questions through fair tests or comparisons
- finding patterns in observations
- classifying
- applying ideas in new situations
- finding out information using books, charts, etc.
- communicating information in various ways, by drawing, speaking, writing, etc.
- using simple measuring instruments

For the later phase – in addition:
- defining questions which can be answered by experiment
- proposing hypotheses
- identifying and controlling variables in carrying out investigations
- recording observations systematically
- using evidence critically and logically
- making measurements with appropriate accuracy
- communicating information in the most appropriate form

Scientific attitudes
For both earlier and later phases:
- curiosity
- willingness to put forward ideas
- co-operation
- perseverance
- open-mindedness
- care in handling living things

For the later phase – in addition:
- responsibility in carrying through an activity
- honesty in reporting results

127

- independence in thinking
- self-criticism

Concepts

For both earlier and later phases:
- about living things (needs, characteristics, variety, animal as including ourselves, plant, food, growth, development, life cycle, senses, health)
- about materials (variety and characteristics of the main groups)
- about change (in the sky, weather, living things, materials when put in water – sinking/floating, dissolving/not dissolving)
- about movement (causes, speed)
- about length
- about area
- about volume and capacity
- about mass
- about time
- about cause and effect

For the later phase – in addition:
- about basic life processes (reproduction, growth, feeding, respiration, sensitivity, movement and support, the variety in how these are carried out)
- about interdependence of living things
- about adaptation of living things
- about air (existence, use by living things, inclusion of water vapour, pollution)
- about water (occurrence, use by living things, solvent power, existence as solid, liquid and vapour, pollution)
- about the soil (composition, function, fertility)
- about the earth in the solar system (sun, moon, stars and planets)
- about forces and movement (sources and effects of force, stopping and starting movement, uniform and changing speed)
- about measurement (using arbitrary or agreed units, estimation, approximation)

Further reading

Broadfoot, P. (ed.) (1986) *Profiles and Records of Achievement: A Review of Issues and Practice*, London: Holt, Rinehart and Winston. This book includes contributions from among the most eminent writers on profiling. It deals well with some of the most important issues arising out of profiling.

Broadfoot, P. (1987) *Introducing Profiling: A Practical Manual*, Hong Kong: Macmillan Education. This book discusses well the variety of profiles that are in use, employing a wide range of illustrated examples.

Garforth, D. (1983) *Profile Assessment: Recording Student Progress. A School-Focused INSET Workshop Manual*, Dorchester: Dorset LEA. This is a very useful and succinct manual for school-focused assessment and profiling. The reader is taken through the problems that need to be tackled by teachers embarking on profiling, allowing them to choose the direction they wish to take.

Garforth, D. and Macintosh, H. (1986) *Profiling: A Users Manual*, Oxford: Stanley Thornes. Like the above manual, this book discusses the approach that needs to be taken in order to profile. Here, though, some of the more theoretical aspects of profiling are also included.

Hitchcock, G. (1987) *Profiles and Profiling: A Practical Introduction*, Singapore: Longman. Hitchcock tackles the many wide-ranging issues involved in profiling in an easily readable manner.

Kempa, R. (1986) *Assessment in Science*, London: Cambridge University Press. The various issues that currently face a teacher in the assessment of science are brought together for consideration in this book.

Woolnough, B. and Allsop, T. (1985) *Practical Work in Science*, London: Cambridge University Press. This book looks at practices in science and the place of practical work. Practical trends are placed in an historical context.

References

AEB (1987) *Candidate Record Sheet. Form 607/3/C*, Guildford: AEB.

AMMA (1983) *Profiles and Records of Achievement: an Introduction to the Debate*, London: AMMA.

Baird, J.A.B. (1981) Industry's knowledge of schools examinations. *Education Research*, **24** (1), 67.

Balogh, J. (1982) *Profile Reports for School-leavers*, pp. 12–36, York: Schools Council/Longman.

Beatty, J.W. and Woolnough, B.E. (1982) Practical work in 11–13 science: the context, type and aims of current practice. *British Educational Research Journal*, **8** (1), 23–30.

Broadfoot, P. (1984) Profiling and the affective curriculum. Paper presented at the Sociology of Education Conference, St Hilda's College, Oxford, 1984, pp. 1–17.

Broadfoot, P. (ed.) (1986) *Profiles and Records of Achievement: A Review of Issues and Practice*, pp. 1–237, London: Holt, Rinehart and Winston.

Brown, S. (1980) What do they know? A review of criterion-referenced assessment. *SED Occasional Papers*, London: HMSO.

Bryce, T.G.K., McCall, J., MacGregor, J., Weston, R.J. and Robertson, I.J. (1983) *Techniques for the Assessment of Practical Skills in Foundation Science. Teachers Guide*, pp. 1–30, Glasgow: Heinemann Education Books.

Bryce, T.G.K. and Robertson, I.J. (1985) What can they do? A review of practical assessment in Science. *Studies in Science Education*, **12**, 1–24.

Burgess, T. and Adams, E. (1980) *Outcomes of Education*, London: Macmillan.

Child, D. (1976) *Psychology and the Teacher*, pp. 310–311, London: Holt, Rinehart and Winston.

City and Guilds (1983) An evaluation of a basic abilities profiling system across a range of education and training provision. *Report for CGLI Profiling Project*, 6.

Cornwall, K.F. (1981) Some trends in pupils evaluation: the growing importance of the teacher's role. *Remedial Education*, **16** (4), 157–61.

CPVE (1984) Consultative document and responses. *The Joint Board of Pre-Vocational Education, 1984.*

Deale, R.N. (1975) Assessment and testing. In *The Secondary School Council Examinations Bulletin*, **32**, 19–27, London: Evans/Methuen Educational.

DES (1982) *Profiles: a review of issues and practice in the use and development of student profiles*, pp. 4–64, Hayes: FEU/DES.

DES (1984) *Records of Achievement: A Statement of Policy*, pp. 1–11, London: HMSO.

DES (1985a) *Science 5–16: A Statement of Policy*, p. 5, London: HMSO.

DES (1985b) *GCSE – The new exam system at 16-plus*, London: HMSO.

DES (1987) *Science Working Group Interim Report*, pp. 70–2, London: HMSO.

DES (1988) *National Curriculum Task Group on Assessment and Testing: A Report*, London: HMSO.

DES/APU (1984a) *Science in Schools: Ages 13 and 15. Research Report No. 3*, London: HMSO.

DES/APU (1984b) *Science: Assessment Framework age 11. Science Report for Teachers: 4*, pp. 6–37, London: HMSO.

DES/APU (1984c) *Science: Assessment Framework Age 13 and 15.* Science Report for Teachers: 2, p. 29–30, London: HMSO.

DES/APU (1985a) *Practical Testing at Ages 11, 13 and 15. Science Report for Teachers: 6*, London: HMSO.

DES/APU (1985b) *Science at Age 13 and 15: Sample Questions*, pp. 142–44, London: HMSO.

DES/APU (1987) *Assessing Investigations at Ages 13 and 15, Science Report for Teachers: 9*, pp. 17, 30, London: HMSO.

Doran, R.C. (1978) Assessing the outcomes of science laboratory activities. *Science Education*, **62** (3), 401–9.

Dore, R. (1976) *The Diploma Disease*, London: Unwin Education Books.

Ebel, E.L. (1979) *Essentials of Educational Measurement*, New Jersey: Prentice Hall.

Eglen, J.R. and Kempa, R.F. (1974) Assessing manipulative skills in practical chemistry. *The School Science Review*, **56**, 261–73.

Frith, D.S. and Macintosh, H.G. (1984) *A Teachers' Guide to Assessment*, pp. 7–17, Glasgow: Stanley Thornes.

Gagne, R.M. (1970) *The Conditions of Learning*, New York: Holt, Rinehart and Winston.

Garforth, D. (1983) *Profile Assessment: Recording Student Progress. A School-Focused INSET Workshop Manual*, pp. 26–99, Dorchester: Dorset LEA.

Garforth, D. and Macintosh, H. (1986) *Profiling: A User's Manual*, pp. 111–34, Oxford: Stanley Thornes.

Gerlach, V.S. and Sullivan, H.J. (1969) *Objectives, Education and Improved Learner Achievement, in Instructional Objectives*, Chicago: Round McNally, 1969.

Goacher, B. (1983) *Recording Achievement at 16 plus*, York: Longman/ Schools Council.

Gunning, D.J. (1978) Report on the assessment of practical work in science subjects on the ordinary grade. *Dalkeith, Scottish Examination Board*.

Gunning, D.J. and Johnstone, A.H. (1976) Practical Work in the Scottish O-grade. *Education in Chemistry*, **13**, 12–14.

Habermas, J. (1976) *Legitimation Crisis*, London: Heinemann.

Hargreaves, A. (1986) *Record Breakers*, p. 203–6, London: Holt, Rinehart and Winston.

Harlen, W. (1978) Evaluation and the teacher's role. *School Council Research Study*, London: Macmillan Education.

Harlen, W. (1983) *Guides to Assessment in Education*, pp. 7–10, Hong Kong: Macmillan.

Harlen, W. (1984) The impact of APU science work at LEA and school levels. *Journal of Curriculum Studies*, **16** (1), 89–102.

Hilton, B. (1984) Profiling skills in science. *Education in Science*, 32–4.

Hirst, P.H. (1974) *Knowledge and the Curriculum*, London: Routledge and Kegan Paul.

HMI/DES (1970) *Curriculum 11–16*, p. 10, London: HMSO.

Hodson, D. and Brewster, J. (1985) Towards science profiles. *The School Science Review*, **67** (239, December), 231–40.

Hofstein, A. and Lunetta, V.N. (1982) The role of the laboratory in science teaching: neglected aspects of research. *Review of Educational Research*, **52** (2), 201–17.

ILEA (1984) *Improving Secondary Schools: the Report of the Hargreaves Committee*, London: ILEA.

IPM (1984) *Schools and the World of Work: What do Employers Look for in School Leavers?* London: IPM.

Jeffrey, J.C. (1967) Evaluation of science laboratory instruction. *Science Education*, **51**, 186–94.

JMB (1978) *Instructions and Guidance for Teachers on the Internal Assessment of Practical Skills in Biology (Advanced) (Biol 5)*, Manchester: JMB.

JMB (1982) *Nuffield Biology (Advanced) Syllabus*, pp. 4–7, Manchester: JMB.

Jones, J. (1983) *The Use Employers Make of Examination Results and other Tests for Selection and Employment: A Criterion Report for Employers*, Reading: School of Education/University of Reading.

Kelly, P.J. (1971) Evaluating Studies of the Nuffield A-Level Biology Trials. *Journal of Biological Education*, **5**, 315–27.

Kelly, P.J. and Lister, R.E. (1969) Assessing practical ability in Nuffield A-Level biology. In J.F. Eggleston and J.F. Kerr (eds) *Studies in Assessment*, London: English Universities Press.

Kempa, R.F. and Ward, J.E. (1975) The effect of different modes of task orientation on observational attainment in practical chemistry. *Journal of Research in Science Teaching*, **12**, 69–76.

Kirkham, S. (1985) 14–18 Pre-vocational education. *The Times Educational Supplement*, 11 January, 10–11.

Law, B. (1984) *Uses and Abuses of Profiling: A Handbook on Reviewing and Recording Student Experience and Achievement*, London: Harper and Row.

Lynch, P.P. (1978) High school students' experience of experimental work in physical science and its relation to pupil attainment. *Journal of Research in Teaching*, **15** (6), 543–9.

Lynch, P.P. (1984) The status of practical skills in Tasmanian (Australian) science teaching: some considerations based on recent survey data. *The Australian Science Teachers Journal*, **30**(1), 29–37.

Lynch, P.P. and Ndyetabura, V.L. (1983) Practical work in schools: an examination of teachers' stated aims and the influence of practical work according to students. *Journal of Research in Science Teaching*, **20**(7), 663–71.

Macintosh, H.G. and Hale, D.E. (1976) *Assessment and the Secondary School Teacher*, London: Routledge and Kegan Paul.

Mathews, J.C. (1974) The assessment of attitudes. Techniques and problems of assessment. *Techniques and Problems of Assessment*, London: Arnold.

MEG (1986) *Biology GCSE Examination Syllabuses 1988*, pp. 22–8, Cambridge: MEG.

Ministry of Education (1960) *Secondary School Examinations other than the GCE*. Beloe Report, pp. 20–8, London: HMSO.

Mortimore, J. (ed.) (1984) *Profiles in Action*, pp. 82–6, Hayes: FEU/DES.

NAS/UWT, *Pupil Profiles: Policy Statement*, p. 2, Birmingham: NAS/UWT.

NUT (1983) *Pupil Profiles: A Discussion Document*, p. 3, London: NUT.

OCEA (1987) *Science Teacher Guide*, p. 7, Oxford: Oxford International Assessment Services.

Pickersgill, D. and Ransden, P. (1987) Sheffield Science Skills Achievement Certificate. *Education in Science*, 36.

Rowntree, D. (1977) *Assessing Students: How Shall we Know Them?* pp. 117–67, London: Harper and Row.

Sands, M.K. and Bishop, P.E. (1984) *Practical Biology: A Guide to Teacher Assessment*, pp. 1–122, London: Bell and Hyman.

Schools Council (1975) *Working Paper 53: The Whole Curriculum 13–16*, pp. 113–4, London: Evans/Methuen.

Schools Council (1981) Records of achievement, *Schools Council Newsletter* (37).

Schools Council (1981) *Working Paper 70: The Practical Curriculum*, London: Methuen.

Scottish Education Department (1979) *Issues in Educational Assessment*, Edinburgh: HMSO.

SEG (1986) *GCSE Biology: 1988 Examination*, pp. 55–9, Guildford: SEG.

Stevenson, M. (1983) Pupil profiles; an alternative to conventional examination? *British Journal of Educational Studies*, **31**(2), 102–4.

Swales, T. (1979) *Record of Personal Achievement: An Independent Evaluation of the Swindon RPA Scheme*, London: Schools Council.

Name Index

SUBJECT INDEX